Designing Hybrid Nanoparticles
(Second Edition)

Designing Hybrid Nanoparticles (Second Edition)

Maria Benelmekki
Nanomaterials Lab. College of Engineering, Swansea University, Swansea, UK

IOP Publishing, Bristol, UK

ISBN 978-0-7503-3587-4 (ebook)
ISBN 978-0-7503-3585-0 (print)
ISBN 978-0-7503-3588-1 (myPrint)
ISBN 978-0-7503-3586-7 (mobi)

DOI 10.1088/978-0-7503-3587-4

Version: 20210501

IOP ebooks

British Library Cataloguing-in-Publication Data: A catalogue record for this book is available from the British Library.

Published by IOP Publishing, wholly owned by The Institute of Physics, London

IOP Publishing, Temple Circus, Temple Way, Bristol, BS1 6HG, UK

US Office: IOP Publishing, Inc., 190 North Independence Mall West, Suite 601, Philadelphia, PA 19106, USA

To my family, and many friends and colleagues

'The ability to build something from the most fundamental constituents is a massive breakthrough'

—Robert J Sawyer

Contents

Preface 2nd edition

I am pleased with the publication of this second edition by IOP Publishing. The first edition was published five years ago and, since then, the field of nanoparticles and their hybrid nanocomposites produced by physical methods, has continued to expand at a fast pace. This expanding tendency will continue to grow in the future. Hence, the main objective of this new edition is to update the reader on the new development in the design and characterization of hybrid nanoparticles (HNPs) produced by magnetron-sputtering-based inert-gas-condensation (MS-IGC) method. The new edition retains the scope and the characteristics of the first edition to give a concise overview on the technical approaches of the synthesis and characterization of HNPs, to provide new insights into one of the most promising 'bottom-up' techniques, based on a practical MS-IGC method, and to evaluate the performances of this method under different conditions.

This new edition features updates in chapters 1, 2, 3, 4 and 5. Updates on nanomaterial classification and characterization methods at nanoscale level are provided in chapter 1. In chapter 2, updates on the modified MS-IGC system are presented, including newly developed systems to improve the production yield. In chapter 3, examples of binary NPs deposited with a newly designed cluster-beam system are presented. Chapter 4 includes a new example of the trimetallic NPs deposited in gas-phase. A new chapter (introduced as chapter 5) is added to this edition to discuss how innovative designs of cluster beam sources are contributing to the shape transformation of NPs to form nanorods, nanowires, and branched nanostructures. Chapter 5 (becoming chapter 6) is expanded to discuss, other aspects related to safety in the lab when using the MS-IGC system.

Finally, I would like to highlight that rather than a concise review on new developments and methods of synthesis of nanoparticles in a general way, the emphasis of this book is strictly on the synthesis and characterization of multi-component nanoparticles produced in gas phase, using different versions of inert gas condensation systems.

Preface 1st edition

The design of complex hybrid nanoparticles (HNPs) with advanced functionalities, and the study of their fundamental properties have a major role in the development of a new generation of nanostructured materials. The possibility of tailoring the dimension, composition, and structure of HNPs, represents a major milestone in the control of their unique physico-chemical properties. These properties, combined with the ability to produce high quality HNPs in fairly large amounts at reasonable cost, led to their potential applicability from materials science to nanomedicine. Within the field of nanomedicine, the amazing properties of HNPs, and the possibility to integrate diagnostic and therapeutic entities onto one nanoparticle have made them an interesting tool in many applications such as multimodal bioimaging and targeted drug delivery.

In the last few years, several 'bottom-up' and 'top-down' synthesis routes have been developed to produce tailored HNPs. This book provides a new insight into one of the most promising 'bottom-up' techniques, based on a practical magnetron-sputtering inert-gas-condensation method. A modified magnetron-sputtering-based inert-gas-condensation (MS-IGC) system is presented, and its performance under different conditions is evaluated.

Designed for graduate students, researchers in physics, materials science, biophysics and related fields, and process engineers, this new resource fills a critical need to understand the fundamentals behind the design and tailoring of the nanoparticles produced by MS-IGC method. It shows that the morphology, the size and the properties of the nanoparticles can be modulated by tuning the deposition parameters such as the energy, the cooling rate, and the collision and coalescence processes experienced by the nanoparticles during their formation. The mechanisms of formation of different HNPs are suggested, combining the physico-chemical properties of the materials with the experimental conditions.

This book illustrates the potential of MS-IGC method to synthesize multifunctional nanoparticles and nanocomposites with accurate control on their morphology and structure. However, for a better understanding of HNPs formation, further improvements in characterisation methods of aggregation zone conditions are needed. In addition, the optimisation of the yield and harvesting process of HNPs is essential to make this method sufficiently attractive for large-scale production.

Acknowledgments

I would like to thank Dr E Xuriguera, Dr R E Diaz, Mr T Sasaki, and Mr A Bright for their technical support at TEM; Mr K Baughman and Dr J Vernieres for their technical support at SEM; Dr V Singh for AFM images; Dr A Roberts for his introduction to the Kratos Ultra system; Professor Sowwan (OIST) for providing experimental facilities; and Dr L M Martinez for the technical support and useful commentary for magnetic field mapping. I am grateful to the reviewers of this edition for their useful comments and suggestions.

Author biography

Maria Benelmekki

Maria Benelmekki (Honorary Senior Research Fellow, SWAN-UK). After completing her PhD in solid state physics (UAB-Spain) and postdoctoral work (ENSAM-France), she joined industrial research as Project Manager and then as Project leader, working in R&D international projects related to a broad range of nanomaterials and their applicability in food packaging and automotive industries. She was awarded the REDFUE price 'University-company 2002'. Eight years later, she was awarded a FCT Research Professor Fellowship, and thus joined the Centre of Physics of UM-Portugal, where she built up a research topic in the area of nanoparticles and hybrid nanomaterials. She extended her research to fundamental aspects of materials surfaces and interfaces and their performance for biomagnetic separation. In 2012, she moved to Japan where she worked as Senior Staff Scientist at OIST. From 2016 to 2018 she served as Associate Professor at the Department of Materials Science and Engineering, NTNU-Norway. Currently, she is leading an industrial project for the implementation of a patented magneto-optical method for accurate detection and separation of small biomolecules in physiological fluids.

Technical abbreviations

AFM	Atomic force microscopy
bPEI	Branched polyethylenimine
BSE	Backscattered electron
CNS-NP	Carbon nanoscrolls-nanoparticles
DLVO	Derjaguin, Landau, Verwey, and Overbeek
EDX	Energy-dispersive x-ray spectroscopy
EELS	Electron energy loss spectroscopy
FFT	Fast Fourier transformation
HNPs	Hybrid nanoparticles
HRTEM	High resolution transmission electron microscope
MACS	Matrix assembly cluster source
MPNPs	Magneto-plasmonic nanoparticles
MS-IGC	Magnetron-sputtering-based inert-gas-condensation
NCs	Nanoclusters
NPs	Nanoparticles
NRs	Nanorods
PDF	Probability distribution function
CDF	Cumulative distribution function
SE	Secondary electron
SEM	Scanning electron spectroscopy
STEM	Scanning transmission electron microscope
STM	Scanning tunneling microscopy
TEM	Transmission electron microscope
XPS	X-ray photoelectron spectroscopy

Chapter 1

Introduction to nanoparticles and nanotechnology

Nanoparticles (NPs) or (zero-dimensional materials) hold a special place in nanoscience and nanotechnology, not only because of their particular properties resulting from their reduced dimensions, but also because they are promising building blocks for more complex nanostructures in nanotechnology. This chapter gives an overview on NPs and their presence in our daily lives. It provides examples of the use of NPs in nanotechnology to obtain different end-products in different economic activity sectors. In addition, a classification of NPs based on their dimension, morphology and chemical composition is presented. NP uniformity and agglomerations, with a special focus on superparamagnetic and magnetoplasmonic NPs and their nanocomposites are discussed.

1.1 Overview of nanoparticles and nanotechnologies

'Nano' is a prefix used to describe 'one billionth', or 10^{-9}, of something. The concept of nanotechnology was introduced by physics Nobel laureate Richard P Feynman in his famous lecture entitled 'There's plenty of room at the bottom' at the December 1959 meeting of the American Physical Society [1]. Since then, there have been many revolutionary developments in physics, chemistry and biology that have demonstrated Feynman's ideas of manipulating matter at the atomic scale. In 1974, Norio Taniguchi (a professor at the Tokyo University of Science) invented the term 'nanotechnology' to describe extra-high precision and ultra-fine dimensions. He introduced the 'top-down approach' by predicting improvements and miniaturization in integrated circuits, optoelectronic devices, mechanical devices and computer memory devices. Approximately ten years later, K Eric Drexler introduced the 'bottom-up approach' when he discussed the creation of larger objects from their atomic and molecular components as the future of nanotechnology [2].

doi:10.1088/978-0-7503-3587-4ch1

Nanotechnologies are now widely considered to have the potential to bring benefits in areas as diverse as drug development, water decontamination, information and communication technologies, and the production of stronger and lighter materials. Nanotechnologies involve the creation and manipulation of materials at the nanometre scale, either by scaling up from single groups of atoms or by refining or reducing bulk materials [3].

While the development of nanotechnologies is a modern multidisciplinary science involving the fields of physics, chemistry, biology and engineering, the production of NPs, both in nature and by humans, dates from the pre-Christian era. For example, the Romans introduced metals with nanometric dimensions in glass-making; the famous Lycurgus cup (currently exhibited at the British Museum), which displays a different colour depending on whether it is illuminated externally (green) or internally (red), contains NPs of silver and gold [4]. In 1857, Faraday reported the synthesis of colloidal gold (and other metals such as Cu, Zn, Fe and Sn) and its interaction with light [5]. For an overview and chronological table of nanotechnologies, see [6].

Another example of interest is the case of magnetic NPs. Krishnan [7], illustrated the role that magnetic materials play in biology and medicine. In the field of magnetic NPs, a noteworthy pioneering work was published by Blakemore in 1975 [8], where biochemically precipitated magnetite (Fe_3O_4) was found in the tissues of various organisms including bacteria, algae, insects, birds and mammals. Many of these organisms use biogenic magnetite to sense the Earth's magnetic field for orientation and navigation. For more details on the development of magnetic NP synthesis and its presence in biomedicine and biotechnology see [7].

Throughout the last century, the field of colloid science has developed enormously and has been used to produce many materials, including metals, oxides and organic products [9, 10]. One of the first and most easily prepared magnetic colloidal systems was developed by Stephen Papell of the National Aeronautics and Space Administration in the early 1960s [11]. Papell's colloid consisted of finely divided particles of magnetite suspended in paraffin. To prevent particle–particle agglomeration or sedimentation, Papell added oleic acid as a dispersing agent. Subsequently, similar magnetic suspensions have also been synthesized with different nanometre sized particles of pure elements, such as iron, nickel and cobalt, in a wide range of carrier liquids [12, 13].

Ordinary materials, when reduced to the nanoscale, often exhibit novel and unpredictable characteristics such as extraordinary strength, chemical reactivity, electrical conductivity, superparamagnetic behaviour and other characteristics that the same material does not possess at the micro- or macroscale. A huge range of nanomaterials is currently being produced at an industrial scale, while others are being produced at smaller scales as they are still under research and development (table 1.1).

In summary, a number of examples of NPs with new magnetic, catalytic, magneto-optical or optical properties, among others, that differ from those of the bulk materials have been reported in the scientific literature. Size reduction has been found to be the reason behind many of these novel physical and chemical properties,

Table 1.1. List of nanomaterials commercially available [14].

Non-exhaustive list of nanomaterials either used in industry or under investigation.		
Aluminum	Dendrimers	Platinum
Aluminum oxide	Dimethyl siloxide	Polyethylene
Aluminum hydroxide	Dysprosium oxide	Polystyrene
Antimony oxide	Fullerenes	Praseodymium oxide
Antimony pentoxide	Germanium oxide	Rhodium
Barium carbonate	Indium oxide	Samarium oxide
Bismuth oxide	Iron and iron oxides	Silanamine
Boron oxide	Lanthanum oxide	Silicon dioxide
Calcium oxide	Lithium titanate	Silver
Carbon black	Manganese oxide	Carbon nanotubes
Cerium oxide	Molybdenum oxide	Tantalum
Chromium oxide	Nanoclays	Terbium oxide
Cluster diamonds	Neodymium oxide	Titanium dioxide
Cobalt and cobalt oxide	Nickel	Tungsten
Colloidal gold	Niobium	Yttrium oxide
Copper (II) oxide	Palladium	Zinc oxide

Table 1.2. General classification and potential applications of NPs [14].

Product areas with end-products containing NPs	Sectors where nanotechnology is expected to have a considerable impact
• Cosmetics and personal care products • Paints and coatings • Household products • Catalysts and lubricants • Sports products and textiles • Medical and healthcare products • Food and nutritional ingredients • Food packaging and agrochemicals • Veterinary medicines • Construction materials • Consumer electronics	• Medical and pharmaceutical sector • Bio-nanotechnology, bio-sensors • Energy sector, including fuel cells, batteries and photovoltaics • Environment sector including water remediation • Automotive sector • Aeronautics sector • Construction sector, including reinforcement of materials • Composite materials • Electronics and optoelectronics, photonics

which allow a wide range of applications with economic benefits. In 2010, more than 1000 products containing NPs became commercially available (table 1.2) [15, 16].

1.2 Classification of nanomaterials

Typically, NPs are defined as an agglomeration of atoms and molecules in the range of 1–100 nm. They can be composed of one or more species of atoms (or molecules) and can exhibit a wide range of size-dependent properties. Within this size range, NPs bridge the gap between small molecules and bulk materials in terms of energy states [17]. NPs are generally classified based on their dimensionality, morphology, composition, uniformity and agglomeration [18, 19]. However, nowadays, NP classification and categorization are still not well established. In the specific case of hybrid NPs (HNPs), the term 'hybrid' is commonly associated with organic–inorganic NPs and nanocomposites [20]. However, in this book the term 'hybrid' is used according to the Cambridge dictionary definition, where it refers to 'something that is a combination of two different things, so it has qualities relating to both of them' [21].

1.2.1 Dimensionality

- ***Nanoscale in one dimension***. Materials with one dimension in the nanometre scale are typically thin films or surface coatings. Thin films have been developed and used for decades in various fields including electronics, information storage systems, chemical and biological sensors, fibre-optic systems, and magneto-optic and optical devices. Thin films can be deposited by various methods and can be grown controllably at the atomic level (monolayer) [22]. Those nanomaterials are known as 2D materials.
- ***Nanoscale in two dimensions***. Nanomaterials having two dimensions in the nanometre scale. These include for example, nanotubes, dendrimers, nanowires, fibres and fibrils (known as 1D materials). Free particles with a large aspect ratio with dimensions in the nanoscale range are also considered 1D nanomaterials. The properties of 1D materials are less understood and manufacturing capabilities are less advanced.
- ***Nanoscale in three dimensions***. Materials that are nanoscale in all three dimensions are commonly defined as 0D nanomaterials. These include quantum dots or nanocrystals, fullerenes, particles, precipitates, and colloids. Some 0D systems, such as natural nanomaterials and combustion products, metallic oxides, carbon black, titanium oxide (TiO_2) and zinc oxide (ZnO) are well known, while others such as fullerenes, dendrimers and quantum dots represent the greatest challenges in terms of production and understanding of properties.

Figure 1.1 shows examples of nanomaterials with different dimensions. All the samples were deposited on a Si (111) substrate using the magnetron-sputtering-based inert-gas-condensation (MS-IGC) method as described in figures 2.1 and 2.2. The materials shown in figures 1.1(a) and (b) can be classified as 2D nanomaterials, while the Cu NPs shown in figure 1.1(c) are classified as 0D nanomaterials. The iron nanorods shown in figure 1.1(d) can be classified as 1D nanomaterials.

1.2.2 The morphology of NPs and nanocomposites

The morphological characteristics to be taken into account are the flatness, aspect ratio and spatial position of each element in the case of HNPs. A general

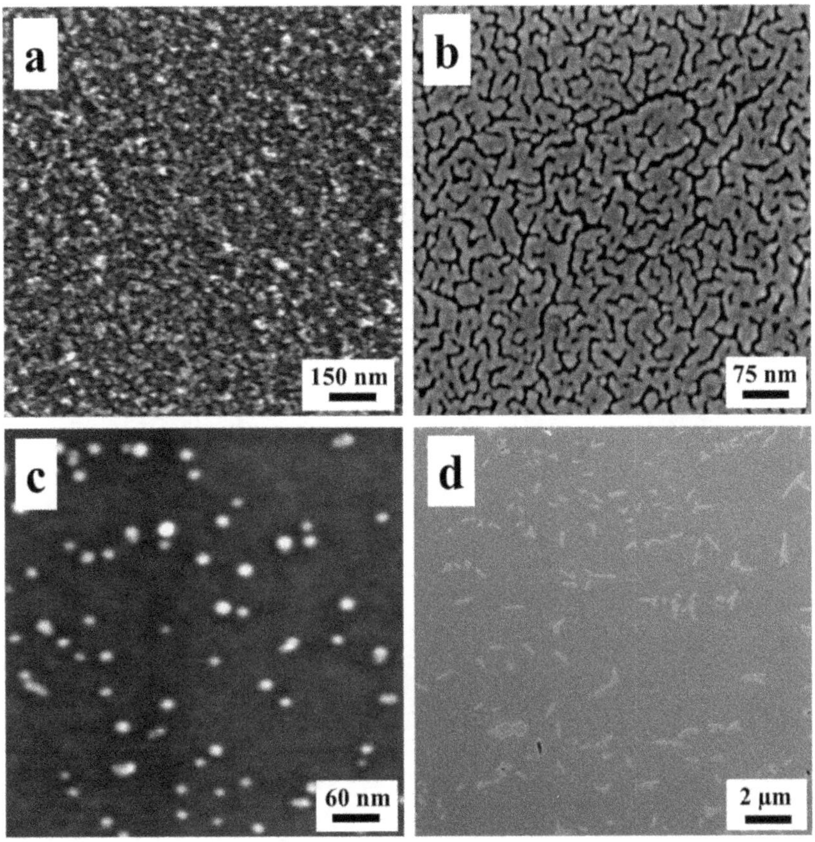

Figure 1.1. Scanning electron microscope (SEM) images showing (a) film of Ti NPs of 80 nm of thickness, (b) near-percolating Au film, (c) monodispersed Cu NPs, and (d) Fe nanorods (NRs).

classification exists between high and low aspect ratio particles. High aspect ratio NPs include nanotubes and nanowires. Low aspect ratio morphologies include spherical, oval, cubic, prism, helical and pillar shapes. Figure 1.2 shows examples of different morphologies of NPs and nanocomposites. Transmission electron microscopy (TEM) images of monodispersed Cu NPs, Fe nanorods and Cu core–Si shell NPs are shown in figures 1.2(a), (b) and (c), respectively. The details of the preparation methods for these NPs are presented in chapter 2. The TEM images in figures 1.2(d) and (e) show a porous magnetite NP and magnetite cubes decorated with Ni nanocrystals, respectively. These NPs were designed and synthesized using the hydrothermal process for purification of histidine-tagged proteins [23].

With regard to nanocomposites, substantial progress has been made in recent years in developing technologies in the fields of magnetic microspheres, magnetic nanospheres and ferrofluids. Nanospheres and microspheres containing a magnetic core embedded in a non-magnetic matrix are used in numerous biological applications [7]. They are used, for example, as carriers that can be targeted to a particular site by using an external magnetic field. In addition, the magnetic separation of

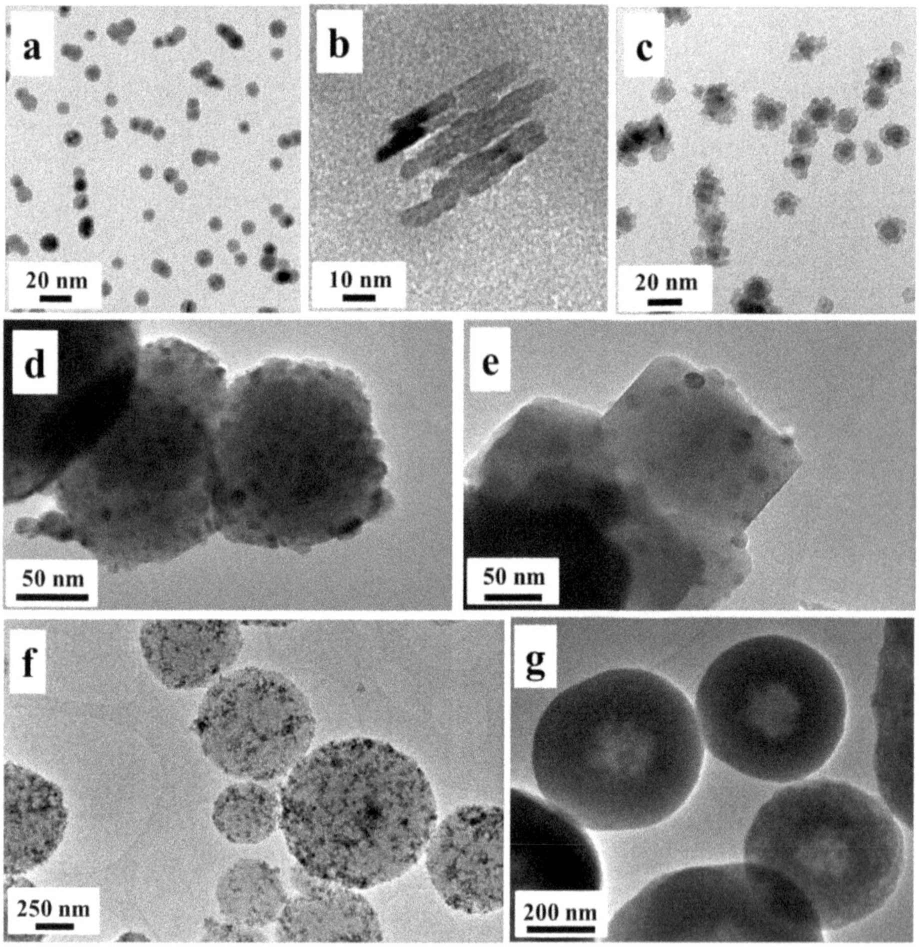

Figure 1.2. Transmission electron microscope (TEM) images of examples of NPs with different morphologies and compositions. (a) Monodispersed Cu NPs, (b) Fe nanorods, (c) Cu–Si core–shell NPs, (d) porous Fe_3O_4 NPs, and (e) Fe_3O_4 cubes decorated with Ni NPs. (f) Porous silica spheres with γ-Fe_2O_3 NPs adsorbed on their surfaces, (g) γ-Fe_2O_3 NPs embedded in porous silica spheres. For more details on the preparation and characterisation of these composites see [24, 25].

organic compounds, proteins, nucleic acids and other biomolecules and cells from complex reaction mixtures is becoming the most suitable method for large scale production in bioindustrial purification and extraction processes. For *in vivo* applications, it is imperative that well-defined biocompatible coatings surround the magnetic particles to prevent any aggregation and also to enable efficient protection of the body from toxicity. However, for *in vitro* applications, biocompatible coatings are not essential; particles can be coated with non-toxic materials inert to chemical and biological media. The particles employed in all these applications are mainly superparamagnetic colloids with appropriate coatings, guaranteeing the stability and biocompatibility of the solutions.

Superparamagnetic NPs exhibit magnetizations of magnitudes similar to those of ferromagnetic materials, however, they have neither coercivity nor remanence. This behaviour, which is of quantum origin, is limited to nanocrystals with sizes below the critical size [26]. Conversely, most applications require superparamagnetic colloidal dispersions with large magnetic responses. Because the magnetization of a particle is proportional to its volume, the maximum magnetization that one can achieve is limited by the critical size of the superparamagnetic transition, which depends on the material [7]. A well-established strategy to create superparamagnetic particles with larger superparamagnetic responses is using nanocomposites (see, for example, figures 1.2(f) and (g)). These superparamagnetic composites are typically made by embedding superparamagnetic nanocrystals in a non-magnetic matrix such as polystyrene or nanoporous silica [27–29]. The resulting colloidal particles retain the superparamagnetic response of their constituent nanocrystals and show larger magnetization when an external magnetic field is applied. Furthermore, neither coercivity nor remanence are observed at the working temperature. However, in addition to the intrinsic superparamagnetic behaviour of the constituent NPs, one must consider the interactions between the NPs inside the skeleton matrix due to their proximity and surface effects due to the coating; these can lead to changes in the overall magnetic response of the colloidal particle.

1.2.3 NP chemical composition

NPs can be composed of a single constituent material or be a composite of several materials. The NPs found in nature are often agglomerations of materials with various compositions, while pure single-composition materials can be easily synthesized using a variety of methods (see chapter 2).

There are three main types of chemical ordering in HNPs (figure 1.3) that describe the way in which the atoms of the elements are arranged within the same NP [30, 31]:

- ***Mixed NPs***: can be either random or ordered (figure 1.3(a)). Randomly mixed alloys correspond to solutions of solids, whereas ordered nanoalloys correspond to ordered arrangements of A and B atoms.

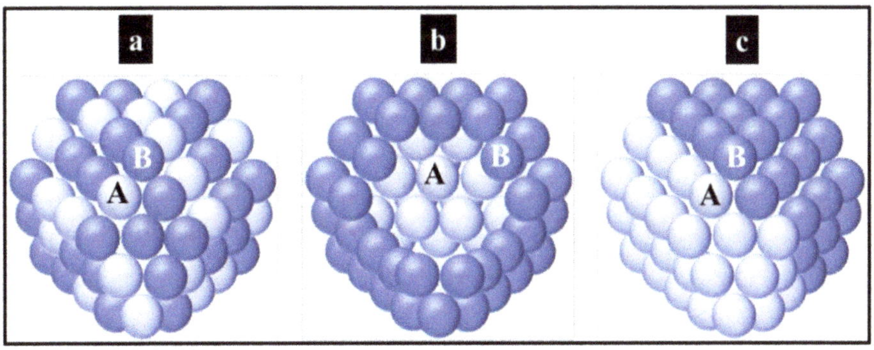

Figure 1.3. Schematic images of binary NPs: mixed structure (a), core–shell structure (b), and layered structure (c) of A and B elements.

- ***core–shell NPs***: consist of a shell of one type of atom (B) surrounding a core of another type of atom (A) (figure 1.3(b)). This pattern is generally denoted by A@B and is common for a large class of NPs. Various thermodynamic considerations, discussed further in chapters 3 and 4, lead to the segregation of materials within the core or shell. A subset of the core–shell category consists of multishell (or 'onion-like') NPs. These NPs have alternating A–B–A shells, or A–B–C in the case of ternary NPs as depicted in figure 1.4(b). The formation of these latter structures is discussed in chapter 4, where it is demonstrated that Fe–Ag–Si multishell structures are obtained by adjusting the experimental conditions.
- ***Layered NPs*** are commonly referred to in the literature as Janus (or 'dumbbell-like') NPs. They consist of two types of NPs (A and B) sharing a common interface (figure 1.3(c)). These types of NPs tend to minimize the number of bonds between elements A and B. This heterojunction structure facilitates phase separation.

Because of the increasing need for multifunctional NPs, other complex structures of NPs such as multicore–shell structures in which the cores can present either 'dumbbell-like' or 'onion-like' structures have been reported in the literature [32–36]. This type of NP is presented and discussed in chapter 4. Another multifunctional subset of the core–shell arrangement is an alloyed single core NP encapsulated in an inert shell (figure 1.4(c)). The case of FeAl@Al_2O_3 NPs [37] is presented and discussed in section 3.2.

TEM images of NPs with various compositions and morphologies are shown in figure 1.5. These NPs were prepared using the MS-IGC method. Control over the composition and morphology of these NPs was achieved by adjusting the experimental conditions as explained in chapters 3 and 4. A detailed description of the deposition system is given in chapter 2.

1.3 NP uniformity and agglomeration

Based on their chemistry and electromagnetic properties, NPs can exist as dispersed aerosols, suspensions/colloids or in an agglomerate state. In fact, NPs tend to adhere

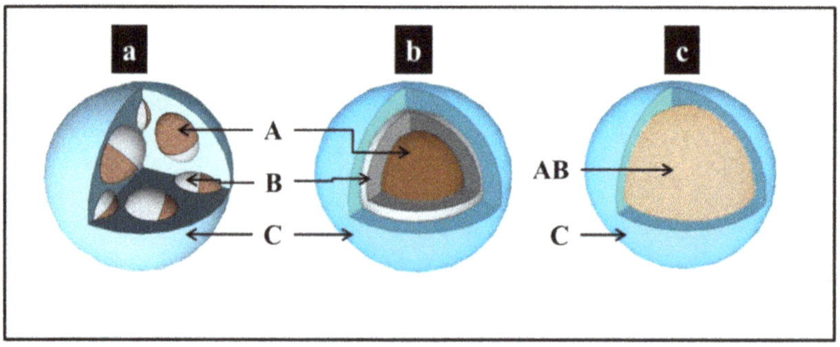

Figure 1.4. Schematic images of ternary NPs formed of elements A, B and C: (a) multicore@shell morphology (the cores present a dumbbell-like morphology). (b) core@multishell morphology, (c) alloyed-core@shell morphology.

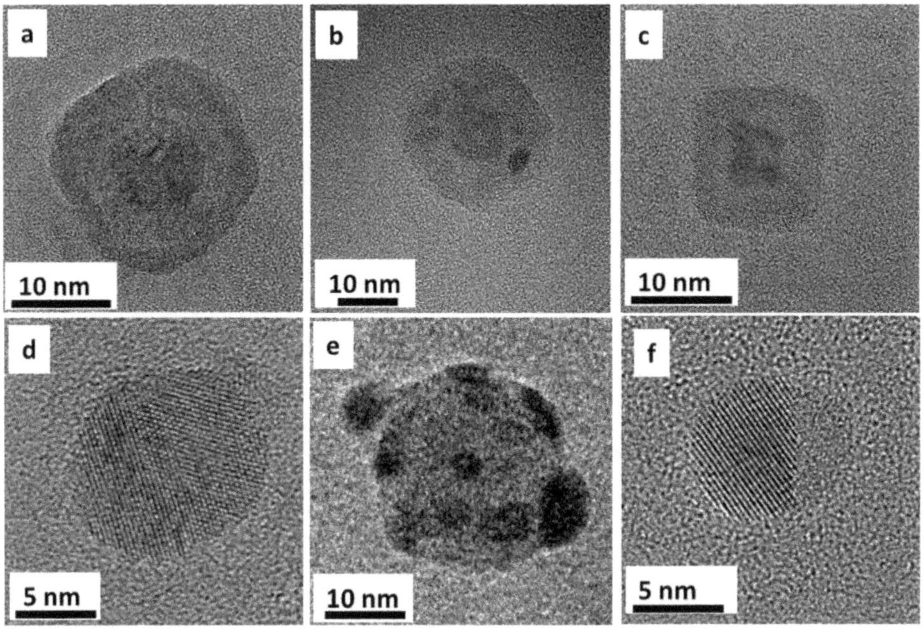

Figure 1.5. (a) Cu@Ag core–shell NP, (b) Cu@Si multicore–shell NP, (c) Fe@Fe_2O_3 core-shell NP, (d) CuAg mixed NP, (e) Si NP inoculated with Ag NPs resulting in a satellite morphology. (f) FeAg dumbbell-like NP; crystalline hemisphere corresponds to Ag.

to each other and to form agglomerates because of van der Waals forces that act over short distances, magnetic interactions, electrostatic forces present in the particles and adhesion forces related to the liquids adsorbed on their surfaces. Agglomeration due to Brownian motion is classified as 'coagulation'.

To avoid agglomeration, several processes include a post-synthesis stage to modify the particle surface by coating it with another organic or inorganic substance. In an agglomerate state, NPs may behave as larger particles, depending on the size of the agglomerate. For example, magnetic NPs tend to cluster, forming an agglomerate state, unless their surfaces are coated with a non-magnetic material. Figure 1.6 illustrates the typical process of stabilization of γ-Fe_2O_3 NPs in an aqueous suspension. The molecules of the anionic surfactant sodium dodecyl sulfate (SDS) are adsorbed onto the surfaces of the nanocrystals providing a negative charge in water. Therefore, the nanocrystals in solution repel each other electrostatically resulting in a stable colloidal suspension. For more details on this procedure see [38]. Other methods commonly used for the stabilization of colloidal NPs consist in their surface functionalization with biological molecules such as oligonucleotides, resulting in strong electrosteric stabilization. In fact the oligonucleotides are negatively charged due to phosphate (PO_2^-) groups along their backbones, inducing a stability based on electrostatic repulsions NP–NP. On the other hand, steric stabilization induced by the anchored oligonucleotides on the surface of the NPs is likely to contribute to the stability of the NPs [39].

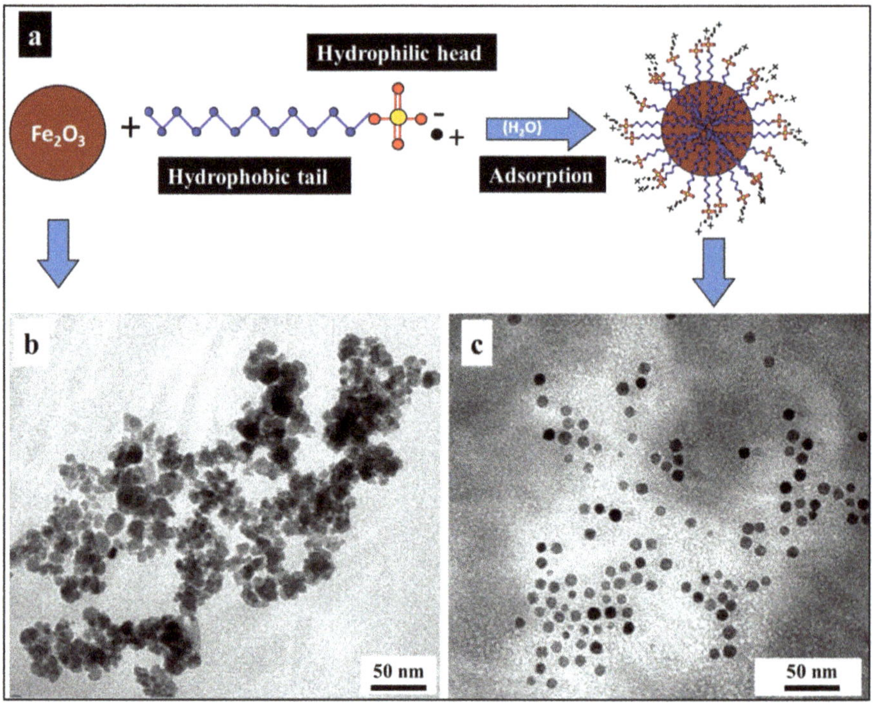

Figure 1.6. (a) Illustration of the stabilization process applied to γ-Fe_2O_3 NPs using SDS (sodium dodecyl sulfate) surfactant. (b) TEM image of the precipitated Fe_2O_3 NPs without SDS. (c) TEM image of SDS-modified Fe_2O_3 NPs (adapted from reference [38] by permission of The Royal Society of Chemistry).

In the case of NPs deposited using vapour phase methods, the NPs are generally deposited on a solid substrate. The transfer of these NPs to a stable suspension is still under investigation. For example, attempts were made to co-deposit NPs from the vapour phase with a beam of water vapour, methanol, or isopropanol onto a nitrogen-cooled substrate [40, 41]. However, the stability of the resulting suspensions was not reported. Recently, a simple and environmentally friendly method for harvesting NPs was developed using polyvinylpyrrolidone (PVP) as a stabilizer [34]. PVP was selected as a non-toxic polymer with good wetting properties. Figure 1.7 shows the procedure used to harvest the NPs to a stable and homogeneous colloidal suspension.

Other physical methods to produce dispersed and stable NPs consist in embedding the NPs in one-dimensional free-standing nanostructures such as ultra-thin films, also known as 'nanosheets' [42], so that the NPs are spatially fixed within the film (or matrix). For example, figure 1.8 shows an example of Fe@Ag@Si onion-like HNPs embedded in an amorphous carbon thin film. These HNPs were deposited using the MS-IGC system described in figure 2.2 [43]. The procedure consisted of the deposition of the HNPs on a Si substrate previously coated with a b-PEI and an amorphous carbon thin films, respectively. For more details see [42, 43].

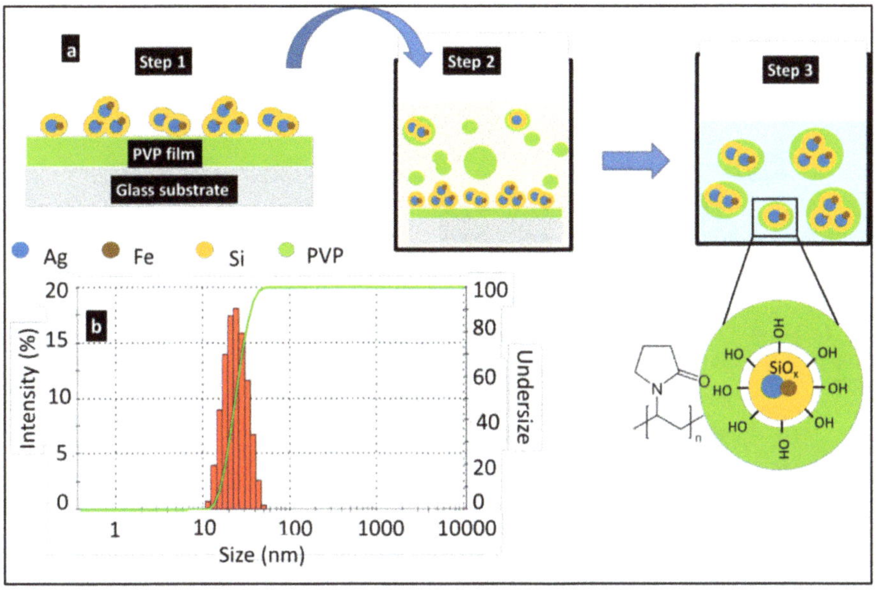

Figure 1.7. (a) Schematic of the exfoliation procedure of the NPs. (Step 1) multicore@shell NPs were deposited on a spin-coated PVP (polyvinylpyrrolidone) film on glass substrate. (Step 2) NPs/PVP/glass samples were immersed in methanol and sonicated for 15 min, then separated to remove excess PVP. (Step 3) After washing precipitated NPs with methanol, they were re-suspended in ultrapure water. (b) Dynamic light scattering histogram showing the size distribution of the hybrid NPs [34]. Reproduced by permission of The Royal Society of Chemistry.

1.4 NP characterization

Once NPs are synthesised, it is important to fully characterise and understand their structure. Over the years, many methods have been developed for a proper understanding of NP structure. In this section, the focus will be on the main techniques with relevance to this book, namely the EM including TEM, SEM and scanning TEM (STEM), electron energy loss spectroscopy (EELS); x-ray photoelectron spectroscopy (XPS); and atomic force microscopy (AFM).

TEM is a very powerful technique for the characterisation of NP size, composition and crystalline structure. When an electron beam interacts with a sample, the electrons can be either transmitted, scattered, backscattered or diffracted [44, 45]. TEM uses the transmitted electron signal to form an image of the sample. The transmitted electron beam is dependent on the sample thickness; for thin samples (a few nanometres), the transmitted electrons pass through without significant energy loss. STEM differs from TEM by focusing the electron beam into a narrow spot that is scanned over the sample in a raster.

Because the attenuation of the electrons depends significantly on the density and thickness of the sample, the transmitted electron beam forms a 2D image of the sample. In hybrid samples, STEM imaging allows the identification of different components based on intensity variation. This intensity variation is related to the

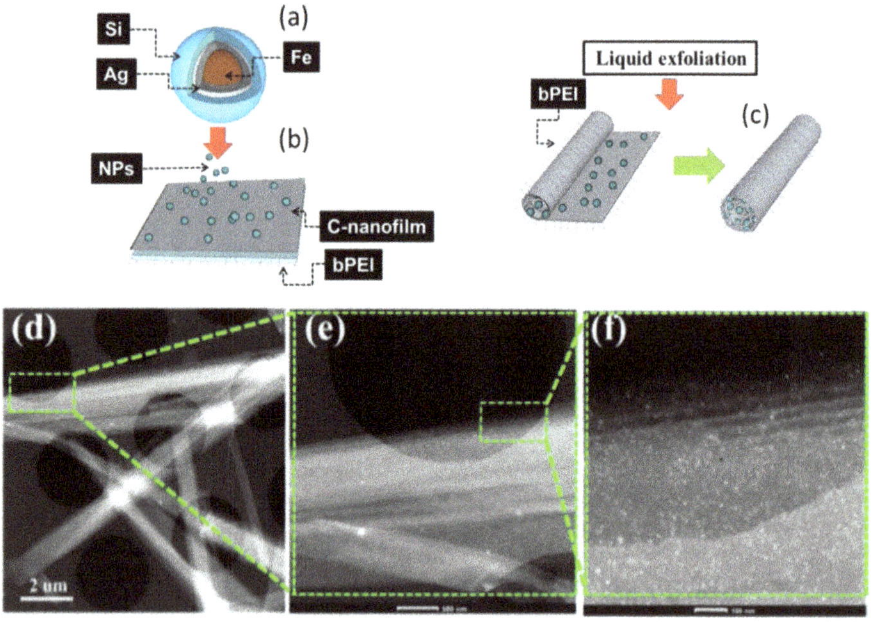

Figure 1.8. (a) Fe@Ag@Si NPs with 'onion-like' morphology are produced through a gas-phase condensation method. (b) NPs are *in situ* deposited on carbon nanofilms supported on bPEI film. (c) After the liquid exfoliation process, a self-scrolling process occurs, resulting in bPEI-functionalized CNS-NP nanocomposites. (d–f) STEM images with different magnifications of the CNS-NPs nanocomposites after the exfoliation process. The bright dots in (f) indicate the NPs embedded in the carbon film. Reproduced with permission from [43]. Copyright 2016 ACS.

difference in the atomic numbers of each component (Z-contrast). In addition, the rastering of the beam across the sample makes it possible to couple STEM with other characterization methods such as EELS [46], allowing direct correlation of image and quantitative data thus obtain details regarding the chemical composition of NPs.

General TEM analysis does not have sufficient resolution to determine the crystallinity of a nanomaterial. However, high resolution TEM (HRTEM) can be successfully employed for the characterisation of the crystallinity of a sample with atomic resolution, as well as for providing information regarding electron diffraction analysis. This approach helps in gaining an insight into the ordering of atoms in an NP. In addition, new generations of EM instruments such as the Titan-ETEM allow performing dynamic studies on the chemical reactions under the effects of the temperature changes and gas pressures at the atomic level. Combined with the image Cs corrector and monochromator technology, this system allows the study of the morphology, structure, composition and bonding of the nanostructures down to the atomic level. Other instruments, such as FEI Talos field emission gun (FEG) TEM, are equipped with highly sensitive Super-X EDS detectors for elemental analysis, and designed to enable the use of gas and liquid operating holders, making possible the study of nucleation and growth nanostructures in both gas and liquid phase.

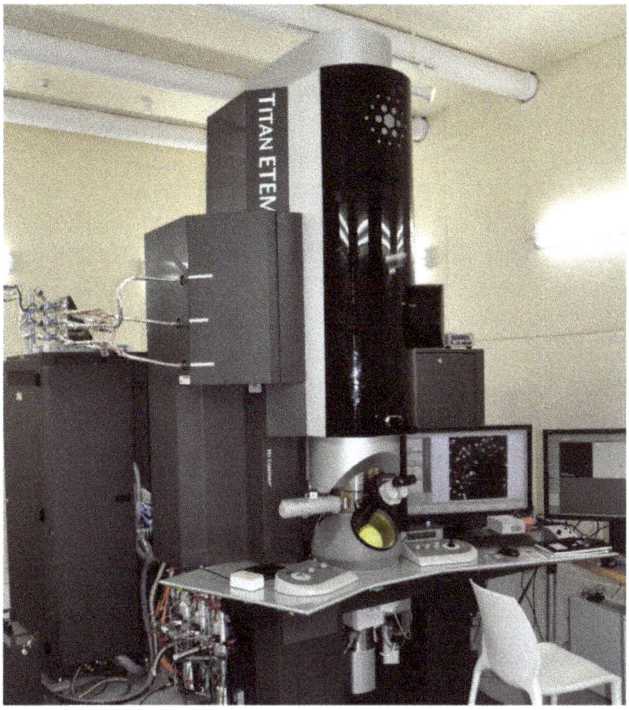

Figure 1.9. TEM used for the characterization of NPs presented in this book. (Credit: OIST-Graduate University).

Figure 1.9 shows a photograph of the TEM used in the characterisation of the NPs presented in this book.

Regarding SEM characterisation, to create an SEM image, a source of electrons is focused into a beam, with a very fine spot (typically the spot size is ~5 nm). The energy of the beam is ranging from a few hundred eV to 50 keV. The incident electron beam is rastered over the surface of the sample. As the electrons strike and penetrate the surface, a number of interactions occur resulting in the emission of electrons and photons from the sample. Three types of SEM images are generated by collecting the emitted electrons and photons: secondary electron (SE) images; backscattered electron (BSK) images; and elemental x-ray maps (EDX). The resolution of modern scanning electron microscopes is typically <10 nm, and image resolution of about 0.5 nm can be achieved in the most recent generation field-emission-gun SEM.

While SEM allows the analysis of conducting and semiconducting samples, AFM is a versatile technique that works for almost any type of surface, encompassing composites, ceramics, oxides, polymers, and biological specimens. This technique was introduced by Binning and Quate in 1986 [47] as an alternative to the earlier developed scanning tunneling microscopy (STM), which was limited to topographic imaging of the surface of conducting samples [48, 49]. Part of the scanning probe microscopy (SPM) family, AFM instrument consists of a nanoscale tip which is scanned over a sample surface. The tip is placed on one side of a soft cantilever, while the opposite side is mounted on a positioning sample stage, which is usually made of an array of piezoelectrics, allowing the control of the distance between the tip and the sample. A laser is pointing at the end of the cantilever, and a mirror and a photodiode collecting the reflected laser beam. The main idea is to make use of a small force taking place between the tip and the atoms at the sample surface. This force depends on the distance between the tip and the sample, and causes the bending of the cantilever. The images are generated by scanning the tip on the surface of the sample, however, instead of adjusting the height of the tip to maintain a constant distance between the tip and the surface as in STM technique, the AFM measures the upward and downward deflections of the tip cantilever while maintaining a constant force of contact [50]. To keep the cantilever bending constant, a feedback loop must change the stage position during the scanning process of the sample. Therefore, a topographic image of the sample surface is obtained.

X-ray photoelectron spectroscopy (XPS), is a surface-sensitive quantitative spectroscopic technique that measures the elemental composition, chemical states, and electronic states of the elements within the material. XPS spectra are obtained by irradiating a material with a beam of x-rays while simultaneously measuring the kinetic energy of electrons that escape from the top 0 to 10 nm of the material being analysed. XPS requires high vacuum ($P \sim 10^{-8}$ mbar) or ultra-high vacuum ($P < 10^{-9}$ mbar) conditions. However, when used to analyse NPs, the importance of the coverage of the NPs must be kept in mind: high coverage leads to high-quality

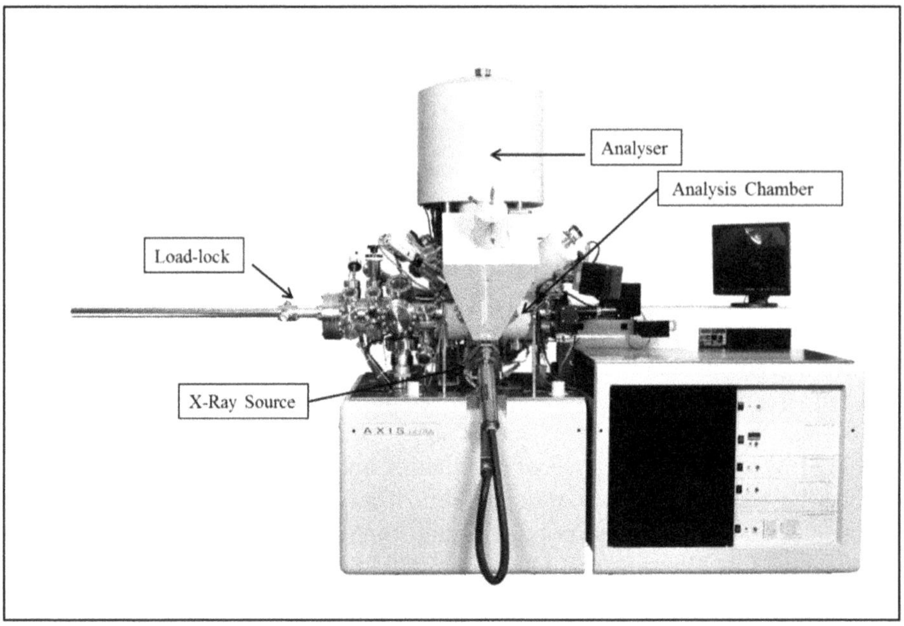

Figure 1.10. X-ray photoelectron spectroscopy system used to analyse the NPs presented in chapters 3 and 4 of this book [51].

spectra. On the other hand, to quantify the composition of the NPs, an inert transfer of the sample to the analysis chamber is necessary to avoid contamination and oxidation of the NP surface. The system used to analyse the NPs presented in this book is shown in figure 1.10. Currently, this system is further developed to a fully automated system under the commercial name Axis-Nova [51].

References

[1] Feynman R P 1960 There's plenty of room at the bottom *Eng. Sci.* **23** 22–36

[2] Drexler K E 1990 *Engines of Creation: The Coming Era of Nanotechnology* (Oxford: Oxford University Press)

[3] www.nano.gov accessed January 2015

[4] http://britishmuseum.org/ accessed January 2015

[5] Faraday M 1857 The Bakerian lecture: experimental relations of gold (and other metals) to light *Phil. Trans. R. Soc. Lond.* **147** 145–81

[6] Satoshi Horikoshi S and Serpone N 2013 *Microwaves in Nanoparticle Synthesis* 1st edn (New York: Wiley-VCH Verlag GmbH & Co. KgaA)

[7] Krishnan K M 2010 Biomedical nanomagnetics: a spin through possibilities in imaging diagnostics and therapy *IEEE Trans. Magn.* **46** 2523–58

[8] Blakemore R 1975 Magnetotactic bacteria *Science* **190** 377–79

[9] Aitken R J, Creely K S and Tran C L 2004 Nanoparticles: an occupational hygiene review HSE http://hse.gov.uk/research/rrpdf/rr274.pdf

[10] Ostiguy C, Lapointe G, Ménard L, Cloutier Y, Trottier M, Boutin M, Antoun M and Normand C 2006 Nanoparticles: actual knowledge about occupational health and safety risks and prevention measures http://irsst.qc.ca/files/documents/pubirsst/r-470.pdf

[11] Papell S 1963 Low viscosity magnetic fluid obtained by the colloidal suspension of magnetic particles US 3215572A

[12] Rosensweig R E 1982 Magnetic fluids *Sci. Am.* **10** 136–45

[13] Kaiser R and Miskolczy G 1970 Magnetic properties of stable dispersions of subdomain magnetite particles *J. Appl. Phys.* **41** 1064–72

[14] JRC Reference Reports 2010 Considerations on a definition of nanomaterial for regulatory purposes https://ec.europa.eu/jrc/sites/default/files/jrc_reference_report_201007_nanomaterials.pdf

[15] Ostiguy C, Roberge B, Woods C and Soucy B 2010 *Engineered Nanoparticles Current Knowledge about OHS Risks and Prevention Measures* 2nd edn (Montréal: IRSST) Report 656

[16] Sebastian V, Arruebo M and Santamaria J 2014 Reaction engineering strategies for the production of inorganic nanomaterials *Small* **10** 835–53

[17] Johnston R L and Wilcoxon J P (ed) 2012 *Frontiers of Nanoscience* (Oxford: Elsevier)

[18] Royal Academy of Engineering and Royal Society 2004 *Nanoscience and Nanotechnologies: Opportunities and Uncertainties* http://nanotec.org.uk/finalReport.htm

[19] Buzea C, Pacheco-Blandino I and Robbie K 2007 Nanomaterials and nanoparticles: sources and toxicity *Biointerphases* **2** MR17–MR172

[20] Gómez-Romero P and Sanchez C 2004 *Functional Hybrid Materials* (Weinheim: Wiley-VCH)

[21] https://dictionary.cambridge.org/dictionary/english/hybrid Accessed November 2020

[22] Seshan K (ed) 2002 *Handbook of Thin-Film Deposition Processes and Techniques - Principles, Methods, Equipment and Applications* (Norwich, NY: William Andrew Publishing/Noyes) pp 1–657

[23] Benelmekki M, Xuriguera E, Caparros C, Corchero J L and Lanceros-Mendez S 2012 Nonionic surfactant assisted hydrothermal growth of porous iron oxide nano-spheres and its application for His-Tagged proteins capturation (unpublished results)

[24] Benelmekki M, Xuriguera E, Caparros C, Rodriguez E, Mendoza R, Corchero J L, Lanceros-Mendez S and Martinez L M 2012 Design and characterization of Ni^{2+} and Co^{2+} decorated porous magnetic silica spheres synthesised by hydrothermal assisted modified-Stöber method for His-tagged proteins separation *J. Colloid Interface Sci.* **365** 156–62

[25] Caparros C, Benlemekki M, Martins P, Xuriguera E, Ribeiro C J, Martinez L M and Lanceros-Mendez S 2012 Hydrothermal assisted synthesis of iron oxide-based magnetic silica spheres and their performance in magnetophoretic water purification *Mater. Chem. Phys.* **135** 510–17

[26] Bean C P and Livingstone J D 1959 Superparamagnetism *J. Appl. Phys.* **30** 120–29

[27] Leun D and Sengupta A K 2000 Preparation and characterization of magnetically active polymeric particles (MAPPs) for complex environmental separations *Environ. Sci. Technol.* **34** 3276–82

[28] Jain T K, Richey J, Strand M, Leslie-Pelecky D L, Flask C A and Labhasetwar V 2008 Magnetic nanoparticles with dual functional properties: Drug delivery and magnetic resonance imaging *Biomaterials* **29** 4012–21

[29] Behrens S 2011 Preparation of functional magnetic nanocomposites and hybrid materials: recent progress and future directions *Nanoscale* **3** 877–92

[30] Paz-Borbon L O 2011 *Computational Analysis of Transition Metal Nanoalloys* (Berlin: Springer)

[31] Tiruvalam R C, Pritchard J C, Dimitratos N, Lopez-Sanchez J A, Edwards J K, Carley A F, Hutchins G J and Kiely C J 2011 Aberration corrected analytical electron microscopy studies of sol-immobilized Au + Pd, Au{Pd} and Pd{Au} catalysts used for benzyl alcohol oxidation and hydrogen peroxide production *Faraday Discuss.* **152** 63–86

[32] Shi W, Zeng H, Sahoo Y, Ohulchanskyy T Y, Ding Y, Wang Z L, Swihart M and Prasad P N 2006 A general approach to binary and ternary hybrid nanocrystals *Nano Lett.* **6** 875–81

[33] Llamosa Perez D, Espinosa A, Martinez L, Roman E, Ballesteros C, Mayoral A, Garcia-Hernandez M and Huttel Y 2013 Thermal diffusion at nanoscale: from CoAu alloy nanoparticles to Co@Au core/shell structures *J. Phys. Chem.* C **117** 3101–8

[34] Benelmekki M, Bohra M, Kim J H, Diaz R E, Vernieres J, Grammatikopoulos P and Sowwan M 2014 Facile single-step synthesis of ternary multicore magneto-plasmonic nanoparticles *Nanoscale* **6** 3532–5

[35] Benelmekki M, Vernieres J, Kim J H, Diaz R E, Grammatikopoulos P and Sowwan M 2015 On the formation of ternary metallic-dielectric multicore-shell nanoparticles by inert-gas condensation method *Mater. Chem. Phys.* **151** 275–81

[36] Benelmekki M and Sowwan M 2015 *Handbook of Nano-Ceramic and Nano-Composite Coatings and Materials* ed H Makhlouf and D Scharnweber (Amsterdam: Elsevier) ch 9 207–23

[37] Vernieres J, Benelmekki M, Kim J H, Grammatikopoulos P, Bobo J F, Diaz R E and Sowwan M 2014 Single-step gas phase synthesis of stable iron aluminide nanoparticles with soft magnetic properties *APL Mater.* **2** 116105

[38] Benelmekki M, Martinez L M, Andreu J S, Camacho J and Faraudo J 2012 Magnetophoresis of colloidal particles in a dispersion of superparamagnetic nanoparticles: theory and experiments *Soft Matter* **8** 6039–47

[39] Benelmekki M, Gasso S and Martinez Ll M 2020 Simultaneous optical and magnetophoretic monitoring of DNA hybridization using superparamagnetic and plasmonic colloids *J. Coll. Surf. Biointefaces* 2020**193** 111126

[40] Haeften K V, Binns C, Brewer A, Crisan O, Howes P B, Lowe M P, Sibbley-Allen C and Thornton S C 2009 A novel approach towards the production of luminescent silicon nanoparticles: sputtering, gas aggregation and co-deposition with H_2O *Eur. Phys. J.* D **52** 11–4

[41] Galinis G, Yazdanfar H, Bayliss M, Watkins M and Haeften K V 2012 Towards biosensing via fluorescent surface site of nanoparticles *J. Nanopart. Res.* **14** 1019

[42] Kim J-H and Benelmekki M 2019 Two-dimensional nanostructures for biomedical applications *Frontiers of Nanoscience* vol 14 (Amsterdam: Elsevier) pp 103–20

[43] Kim J-H and Benelmekki M 2016 Interfacial transformation of an amorphous carbon nanofilm upon Fe@Ag@Si nanoparticle landing and its colloidal nanoscrolls: enhanced nanocompositing-based performance for bioapplications *ACS Appl. Mater. Interfaces* **8** 33121–30

[44] Williams D B and Carter B C 2009 *Transmission Electron Microscopy: A Text Book for Material Science* 2nd edn (Berlin: Springer Ebook Series)

[45] Frenkel A I, Hills C W and Nuzzo R G 2001 A view from the inside: complexity in the atomic scale ordering of supported metal nanoparticles *J. Phys. Chem.* B **105** 12689

[46] Pearson D H, Ahn C C and Fultz B 1993 White lines and d-electron occupancies for the 3rd and 4th transition metals *Phys. Rev.* B **47** 8471–78

[47] Binning G and Quate C F 1986 Atomic force microscope *Phys. Rev. Lett.* **56** 930–33

[48] Binning G, Rohrer H, Gerber C and Weibel E 1982 Tunneling through a controllable vacuum gap *App. Phys. Lett.* **40** 178–80

[49] Binning G, Rohrer H, Gerber C and Weibel E 1982 Surface studies by tunneling microscopy *Phys. Rev. Lett.* **49** 57–61

[50] Cao G and Wang Y 2011 *Nanostructures and Nanomaterials* 2nd edn (Singapore: World Scientific)

[51] www.kratos.com (accessed November 2020)

IOP Publishing

Designing Hybrid Nanoparticles (Second Edition)

Maria Benelmekki

Chapter 2

Production of hybrid nanoparticles

The increasing need for multifunctional nanoparticles (NPs) continues to attract the interest of scientists and engineers and persuades them to develop optimum methods for the production of such challenging nanomaterials. This chapter gives an insight into the novel-configuration of magnetron-sputtering-based inert-gas-condensation (MS-IGC) source as one of the widespread methods for supported hybrid NP production. The modified system allows the production of hybrid NPs (HNPs) with controlled size, morphology and chemical composition, all in a single step. Section 2.1 gives an overview of the methods of production of NPs with a special focus on MS-IGC method. Details on the last modifications to this technique are presented in section 2.2, including insights on the latest improvement in the production yields of the HNPs and the synthesis of anisotropic nanostructures. Finally, a brief description of the factors affecting the formation of the HNPs synthesised by gas-phase methods is given in section 2.3.

2.1 An overview of the production methods for NPs

Different methods of production are used to optimize specific properties of NPs, the yield and the suitability for scaling up. Broadly speaking, methods of NP production can be divided in two main groups: (1) top-down methods and (2) bottom-up methods. Other emerging and promising lithographic procedures, such as e-beam and ion-beam nanolithography and scanning-probe-based lithography, are of high interest in the lab because of the excellent control afforded at the sub-100 nm scale. However, these procedures are not practical for large-scale production due to their high cost and low yields. Nanoprinting (a term commonly used to refer to a range of techniques used to transfer a nanoscale pattern to a suitable surface) can also be used for printing NPs and has the potential to lower the cost considerably compared to other nanofabrication methods. However, these techniques are still at an early stage [1, 2].

doi:10.1088/978-0-7503-3587-4ch2

2.1.1 Top-down processes for the production of NPs

Traditional mechanical grinding methods are able to produce NPs at a relatively low cost from a variety of materials using several milling techniques. This approach is often applied in the production of metallic and ceramic NPs but can also be used to produce complex NPs such as bismuth-telluride-based alloys [3]. Milling involves thermal stress and is energy intensive. Purely mechanical milling can be accompanied by reactive milling in which a chemical or chemo-physical reaction is produced during the milling process. However, the powders obtained show a broad particle size distribution, significant aggregation and impurities from the milling agent.

Physical or chemical exfoliation methods also fall within the top-down category and are often used to obtain laminar nanomaterials such as nanosheets and nanoflakes. In this field, the wide variety of chemical exfoliation methods proposed for the preparation of graphene dispersions should be mentioned. Most are based on the so-called Hummers procedure [4].

2.1.2 Bottom-up processes for the production of NPs

Bottom-up synthesis is used to produce complex nanostructures from atoms to molecules and can often be tailored to obtain good control of the sizes and shapes of NPs. The methods include gas phase and liquid phase reactions. Liquid phase and gas phase synthesis have different time scales; the slower liquid phase processes can be used to obtain thermodynamically controlled products, while for gas phase synthesis kinetic control is often the only option available. Another important distinguishing characteristic between liquid phase and gas phase synthesis is their different behaviour in aggregation: the much higher density and viscosity of the liquid phase slow down NP movement and help to limit aggregation. Moreover, in liquid phase synthesis, NP stability can be further increased through the addition of a stabilizing surface agent.

Gas phase NP production can be divided into methods that operate at atmospheric pressure and those that generate nanocluster (or NP) beams at low pressures [5]. Despite the technical demands and expense of low pressure gas phase methods, they remain a favourite for producing supported NP assemblies. The clusters or NPs can be produced continuously in sizes from 105 atoms per cluster down to molecules. Most NP synthesis methods in this category are based on homogeneous nucleation of a supersaturated vapour and subsequent particle growth by condensation and coalescence [6, 7].

Since their initial development, low pressure gas phase methods have differed in how the elemental vapours are generated: arc discharge, pulsed microplasma, laser vaporization, Joule heating and sputtering have been used [8–11]. In general, the formation of the vapour occurs within a reactor at a relatively low pressure. The precursor materials are introduced into the reactor as solids, powders, liquids or gases, where they are evaporated and mixed with a carrier gas. Supersaturated vapour can be produced by cooling, by chemical reaction or by a combination of both. Next, the nucleation process is initiated by the formation of very small cluster embryos from the molecular phase. These nuclei subsequently grow by surface growth mechanisms (heterogeneous condensation, surface reaction) and by collision and coalescence [12–14].

Subsequently, many groups have used these techniques and methods to produce gas phase alloys and core–shell particles with complete flexibility of choice of the core and shell materials [15]. As the goal of this book is not to review the state-of-the-art of inert-gas-condensation sources for NP production, for more details on this topic see [6, 16–19].

In this book, the focus will be entirely on the MS-IGC source because it allows the production of controlled NPs which are suitable for fundamental studies, such as the intriguing field of oxidation processes at the nanoscale. Additionally, these modified sources allow the production of NPs with tailored chemical compositions, fulfilling the increasing need for multifunctional NPs.

2.2 The MS-IGC method

Of the low pressure gas phase techniques, the MS-IGC method has emerged as one of the most flexible and successful, as it allows control of the growth process by varying parameters such as the magnetron power, the length of the so-called aggregation zone (figure 2.1) and the inert gas pressure. In addition, the sizes of NPs can be monitored and the composition can be altered *in situ* by the introduction of gases such as oxygen and nitrogen to form oxidized and nitrided NPs with tunable stoichiometries. The wide acceptance of magnetron-sputtering may be a consequence of its ability to generate a large proportion of ionized NPs that can be mass/charge selected by attaching a quadrupole to the source (quadrupole aggregation sources became commercially available in 2001) [20, 21].

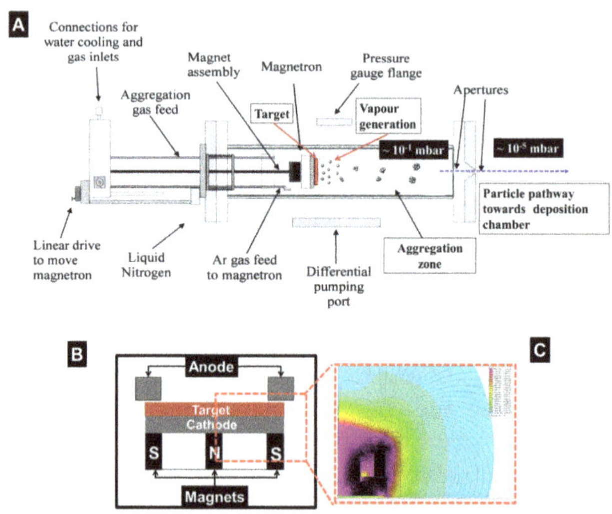

Figure 2.1. (a) A diagram of the aggregation zone of a standard MS-IGC source [20]. (b) The cathode magnetron geometry. (c) The magnetic field configuration in the vertical cross-section of the cathode [21].

The pioneering study published by Haberland and co-workers [11] noted the main advantages of the MS-IGC source as follows. (1) It can produce a very large range of mean NP sizes from 50 to 106 atoms per cluster. (2) It has a high degree of ionization, from 30% to 80%, depending on the target materials, allowing easy acceleration of the NPs. (3) A wide variety of elements and alloys can be used as source materials. (4) The substrate is not in contact with the plasma and the landing temperature can be regulated.

2.2.1 The synthesis of HNPs

An important research direction in NP synthesis is the expansion from single-component NPs to multicomponent hybrid nanostructures with discrete domains of different materials arranged in a controlled fashion. Progress in nanomaterial synthesis has made possible the mixing of two or more different materials to obtain binary, ternary and multicomponent HNPs. The advantages of multicomponent structures lie in three aspects. The first is to realize multifunctionality; different functionalities can be combined in one single NP. The second advantage is in providing novel functions not available in single-component materials or structures. The third advantage is in achieving enhanced properties due to the couplings between the different components. In fact, because many physical and chemical properties have critical length scales on the order of nanometres, the intimate contact between the nanocomponents in HNPs should allow strong interactions between these components and a possible rational modulation of the physical and chemical properties of each individual component.

In the last decade, further progress in gas phase synthesis was made by sputtering alloyed materials to expand NP composition from single components to hybrid heterostructures [22]. For example, Xu *et al* [23] reported experimental work on iron–silver (Fe–Ag) and cobalt–gold (Co–Au) heterostructured NPs with core–shell and dumbbell-like shapes obtained by the sputtering of FeAg and CoAu composite targets, respectively. Wang reviewed and reported on the preparation of mono-dispersed FePt NPs with high anisotropy using the gas phase condensation method with an FePt alloyed sputtering target [24]. He *et al* reported SmCo5 NPs obtained using the gas phase condensation method; they prepared a tailored sputtering target by superposing a Co target over a Sm target. Small holes were drilled in the Co target to obtain simultaneous sputtering of both Co and Sm [25]. For other interesting studies of HNPs prepared using the MS-IGC method from single alloyed sputtering targets such as Co–Au, Mo–Cu and Au–Pd see [26–29].

2.2.2 Modification of the MS-IGC source

Several modifications of gas phase sources have been reported in the literature. Most are aimed at the generation of high cluster fluxes [30–32], post-annealing treatments for the generation of magnetic particles with appropriate crystallographic structure and high magnetic anisotropy [33–35]. Other modifications have addressed the control of cluster oxidation [36], focusing the cluster beam with aerodynamic lenses

[37], control of the cluster landing energy on the substrates [38] and room temperature operation of the source [39], among others.

- *Obtaining multicomponent HNPs*

Nevertheless, only a few works have been reported in the literature on modifications of the NP source to allow the *in situ* control and adjustment of the chemical composition or the position of each element within the generated HNP. For example, Pellarin *et al* used a double ablation laser to generate (C60)nSim cationic clusters by quenching the vapours from two independent C60 and Si targets [40, 41], while Sumiyama *et al* used independent sputtering sources to produce Co–Si and Fe–Si core–shell clusters [42]. In both designs, the positions of the targets were fixed and hence could not be moved to adjust the length of the aggregation zone, as is customary in more standard NP sources [31, 43]. Otherwise, in most cases the chemical compositions of the NPs produced depended on the chemical composition of the single target material. Modification of the chemical composition of the NPs required the substitution of the single target by another with a different chemical composition.

Recently, Martinez *et al* [44] reported on the use of two or more independent target materials to produce AgAuPd alloyed particles using the MS-IGC method. In fact, they modified their system to generate NPs with a controllable and tunable chemical composition while maintaining control of the cluster size. In their patented design [45], they replaced the typical two-inch magnetron with three one-inch independent magnetrons whose powers and positions could be controlled individually.

Alternatively, the Mantis Deposition company [21] has started commercializing the patented Nanogen Trio magnetron sputtering cathode [46], the details of which are presented in figure 2.2. A photograph of the Nonogen Trio showing the three cathodes integrated in a single sputtering head is shown in figure 2.3. This system was used to produce the HNPs presented in chapters 3 and 4 of this book.

These latest modifications opened a new horizon for the design and preparation of HNPs with precise control of their composition, morphology and size, all in a single step. Nevertheless, one of the most challenging tasks when magnetic HNPs are being designed and produced with this novel configuration of magnetrons (figure 2.3) is the efficient sputtering of the magnetic targets. In fact, when a magnetic target is placed on the cathode, it provides a path of magnetic flux generated by the permanent magnet of the cathode. Flux passes through it and only a small portion escapes and reaches the surface of the target to contribute to the generation of magnetron plasma. For this reason, as shown in figure 2.4, a study of the effects of the thickness of the iron target on the magnetic field distribution near the surfaces of the individual magnetrons was performed using a finite-element software package (Maxwell 3d v14). For adequate sputtering, the thickness of the Fe target should not exceed 1 mm.

- *Production of core@shell nanostructures*

Core@shell HNPs are of great importance for various technologies as they exhibit a wide range of advantageous properties, such as the surface passivation and the enhancement of their physical and chemical properties [47]. Different research

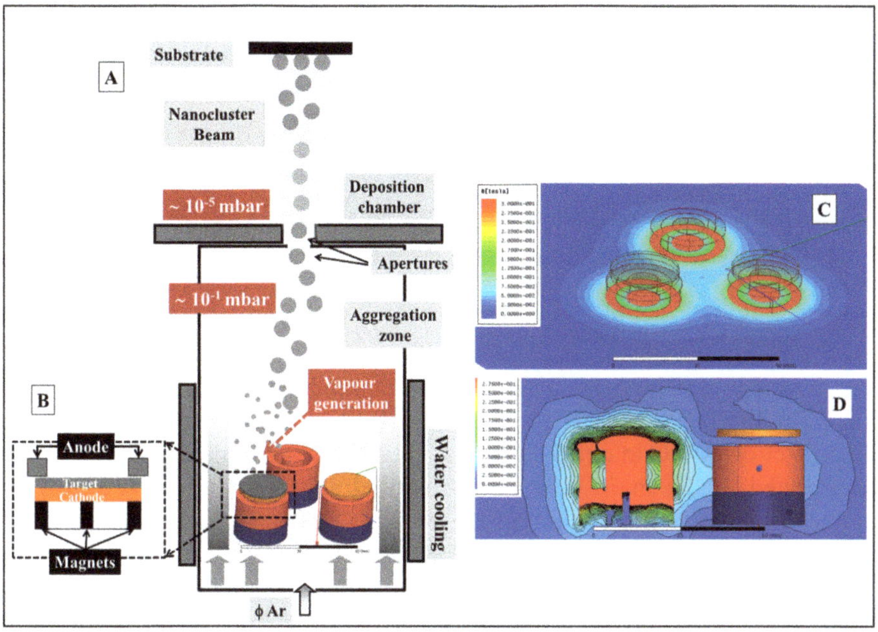

Figure 2.2. (a) A diagram of the modified NP source used to produce the HNPs presented in this work. (b) shows the three independent magnetron-sputtering cathodes; each target is one inch in diameter. The distribution of the magnetic field in the half-height plane crossing the three cathodes of the sputtering trio is plotted in (c). The magnetic field configuration in the vertical cross-section of one cathode is shown in (d). The magnetic field was calculated using finite-element software package Maxwell 3d v14.

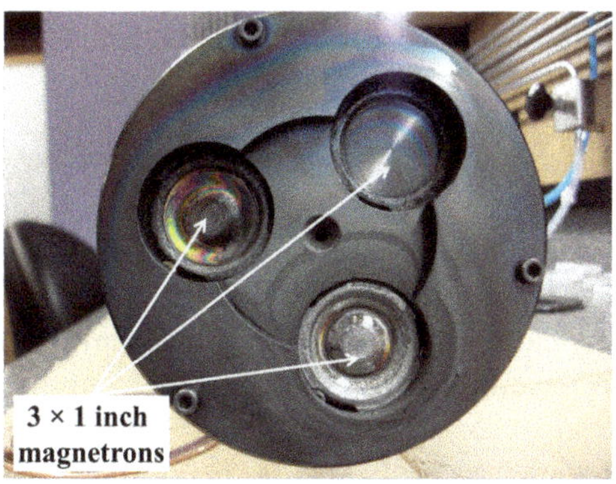

Figure 2.3. A photograph showing the three independent magnetron cathodes integrated in a single sputtering head as shown in figure 2.2 [46].

groups have been working on efficient modification of conventional MS-IGC systems to produce controlled core@shell HNPs. More details can be obtained from reference [47]. Recently, Hanuš *et al* modified their MS-IGC system to successfully produce controlled Ni@Ti core@shell HNPs [48]. The system consisted of a primary gas aggregation chamber where a circular Ni target was DC sputtered

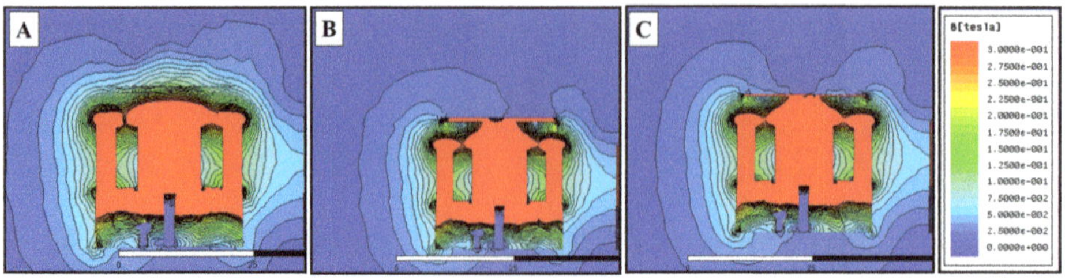

Figure 2.4. The distribution of the magnetic field in the vertical cross-section of the magnetron sputtering cathodes (a) without Fe target, (b) with a 1mm thick Fe target, (c) with a 0.3 mm thick Fe target. The target is one inch in diameter.

for the production of Ni NPs. This primary chamber was attached to a secondary chamber for the coating of Ni NPs with a Ti thin layer. Within this secondary chamber, a circular DC magnetron with a Ti target was installed. The whole system was connected to the main deposition chamber. The targets of Ni and Ti were simultaneously sputtered, so that when the Ni NPs exit the aggregation chamber (aggregation zone), they are in-flight coated with Ti thin film during their pathway towards the deposition chamber [48].

A diagram of a commercial magnetron sputtering shell-coater device is shown in figure 2.5(b) [49]. The shell-coater-chamber is a cylindrical module of external length and diameter of ~295 mm and ~150 mm, respectively. A linear magnetron sputtering cathode of 153 mm length × 51 mm width is placed in the shell coater-chamber as indicated in figure 2.5(b). The shell-coater-chamber is attached adjacently to the aggregation-chamber, with in- and out-apertures of 6 mm of diameter [47, 49]. The walls of the shell-coater-chambers are water-cooled. This system was used by a few research groups to produce core@shell HNPs and nanostructures [50, 51], which will be discussed in chapter 5 of this book.

- *Increasing the production rate of HNPs*

The successful control of size, composition and shape of the HNPs formed by MS-IGS technique has made this method a powerful tool to investigate the fundamental properties of NPs on surfaces and has also stimulated studies for different applications at laboratory levels, catalysis being one of them. In this specific case, producing catalysts for industrial applications is limited by the low production rates of this technique. For example, a conventional MS-IGC system can typically produce about 0.05–0.5 μg Au_{100} NPs per hour, which means that producing 1g of catalysts loaded with 10 mg of NPs (1 wt% loading) for standard catalytic R&D measurements, is far from being reasonable [52]. Therefore, significant scale-up of the NP production rate is crucial for their applications. Within this context, A scale-up of two orders of magnitude has been achieved with a conventional MS-IGC system, by extracting the NP beam as shown in figure 2.6 [53]. The modified instrument consists of four sections: Aggregation-chamber; ion-optics-chamber; mass-filter-chamber, and deposition-chamber. An electrostatic deflector deviates the NP beam down by 90° toward the deposition-chamber (as shown in figure 2.6). Powder loaded into a stainless-steel cup is mechanically

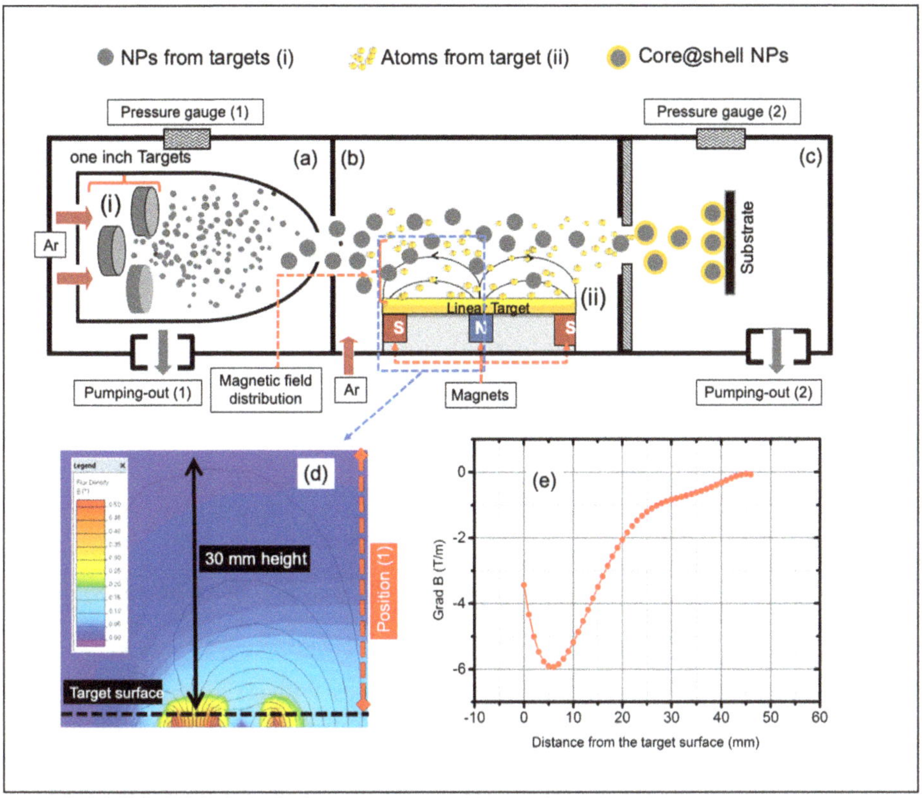

Figure 2.5. A diagram of the modified MS-IGC system (a) Aggregation-chamber containing three one-inch magnetrons; (b) Shell-coater-chamber where the linear-magnetron sputtering target is installed; (c) the deposition-chamber where the HNPs are deposited on the substrates. (d) shows the magnetic field distribution in the vertical cross-section of the linear-cathode (along the short-side of the linear-cathode as indicated in (b)). (e) shows the profile of magnetic field gradient in the region indicated as 'position 1' in (d). These data will be further discussed in chapter 5.

agitated close to the beam deposition to maximize exposure of powder to the incoming NP beam, enabling the deposition of 0.1 wt % Au_{300} clusters onto 1 g of powder in 10 h [52, 53].

In addition, a new type of NP beam source (MACS) has recently been developed in an attempt to reach a production rate of 1 g of catalyst in 1 h [54, 55]. The principle of this new system is different from the traditional method of condensing hot atoms in a cold gas. Instead, the new system consists in the impact of an ion beam (e.g., Ar^+) upon a condensed matrix of rare gas atoms impregnated with atoms of the desired material (figure 2.7(a–b)). The impact of an ion initiates a cascade of collisions inside the matrix, causing clusters to nucleate and grow. A cluster can be ejected from the matrix by the same (or subsequent) ion beam impact. A diagram of the whole modified system is shown in figure 2.7(c). Different results were recently published showing the performances and achievements of such advanced system as for the case of the production of 1g of TiO_2 powder inoculated

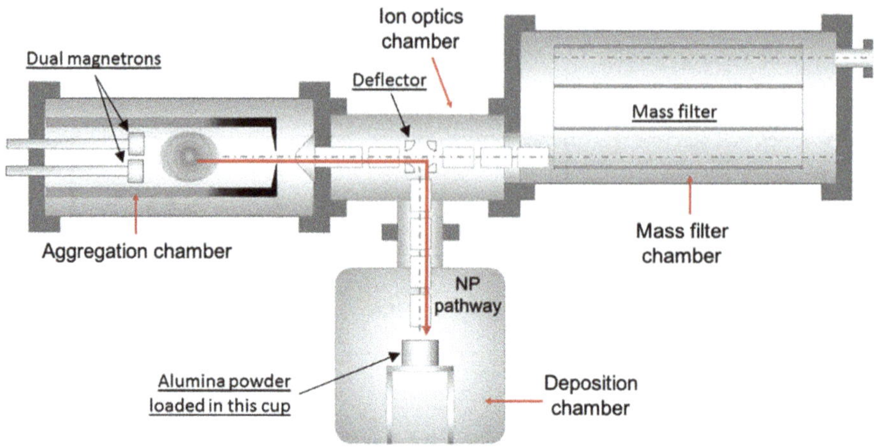

Figure 2.6. A diagram of an MS-IGC system, modified for the deposition of NPs on powders. The NPs are deposited directly onto powders in the deposition-chamber. Adapted from [53] with permission of RSC under CC BY 3.0.

with Ag NPs (figures 2.8) [56]. More details on the performances of this system are given in chapter 3.

- *Self-assembly of HNPs*

The assembly of NPs to produce nanostructures with a large surface area-to-volume ratio, is likely to strengthen nanomaterial applications in different fields [57, 58]. Within this context, solution-based self-assembly of magnetic NPs is the most common practice to produce 1D and 2D nanostructures [51, 59]. However, this method faces limitations such as the surface oxidation of the NPs and the presence of impurities in the final product. Hence, gas-phase methods seem to be an appropriate alternative for the production of pure metallic nanostructures. For example, Ekeroth and co-workers, have developed a gas-phase method to produce magnetic nanowires and branched nanostructures [60, 61], using a modified system, as depicted in figure 2.9. More details on their process and results will be discussed in chapter 5.

Briefly, the system consists of a hollow cathode made of the material to be sputtered, which is placed in a high-vacuum chamber (figures 2.9). The cathode presents an inner diameter of 5 mm and a length of 54 mm. The cathode is mounted on a copper block, which is placed on a magnetron. A cylindrical mesh, with a diameter of 85 mm and a length of 110 mm, is placed in front of the cathode to confine the plasma. The Ar gas is introduced into the vacuum chamber (through the hollow cathode), where the NPs are formed. for more details on the working principle see [60, 61]. To assemble the NPs, a magnetic field is applied by incorporating a magnet to the substrate holder as shown in figure 9.2.

2.3 Factors influencing the formation of HNPs using gas-phase methods

For both fundamental and technological purposes, new developments are strongly dependent on accurate control of the growth of NPs. It is essential to control the

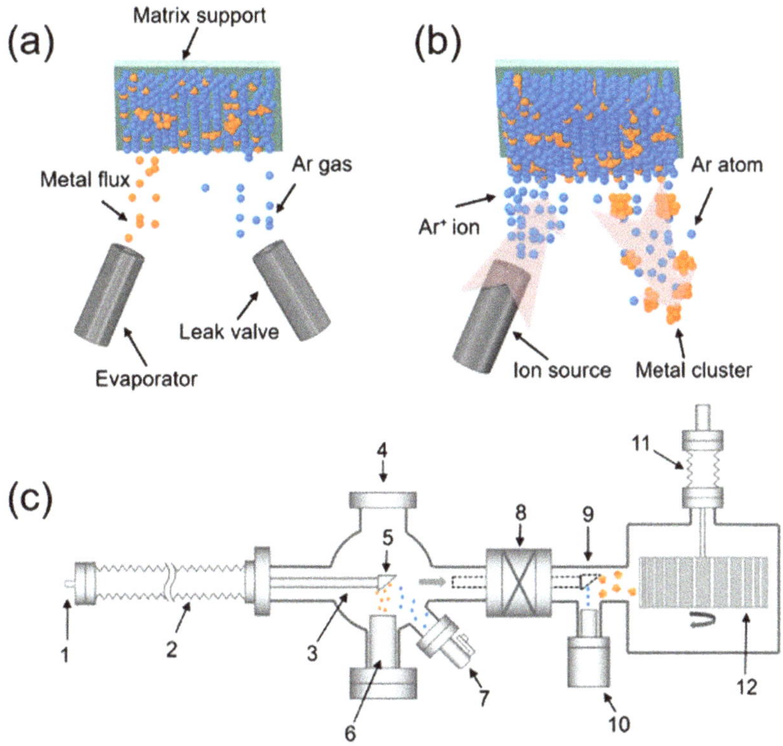

Figure 2.7. Principle of operation of the MACS with (a) formation of matrix by co-condensation of metal and Ar gas onto a cryogenically cooled surface and (b) production of NPs by sputtering of the matrix with Ar ions. (c) Schematic diagram of the MACS system with (1) liquid He and electrical feedthrough for cold finger, (2) linear translator for (3) cold finger, (4) pumping port for the matrix chamber, (5) matrix support, (6) evaporator, (7) Ar leak valve, (8) gate valve, (9) position of the matrix during sputtering to produce nanoclusters, (10) ion source, (11) linear translator, and rotary drive for (12) the sample carousel. Adapted from [55] with permission of AIP under CC BY 4.0.

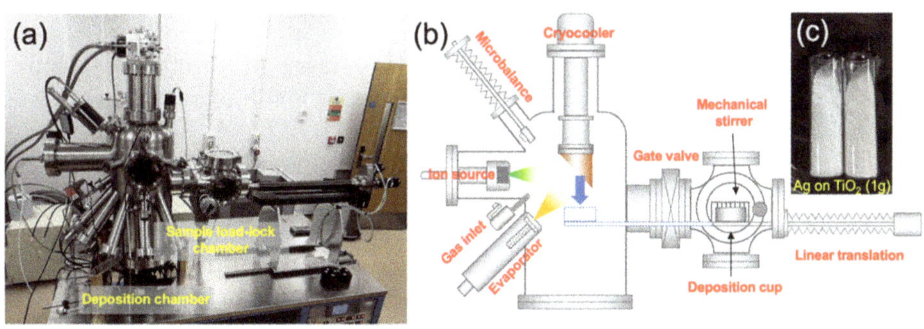

Figure 2.8. MACS system for the preparation of cluster catalysts at the gram scale. (a) Photograph and (b) schematic diagram of the system. (c) Photograph of the Ag cluster on TiO_2 powder catalyst produced (1 g), showing the color change from white to brown. Reproduced from [56] with permission of ACS under CC BY.

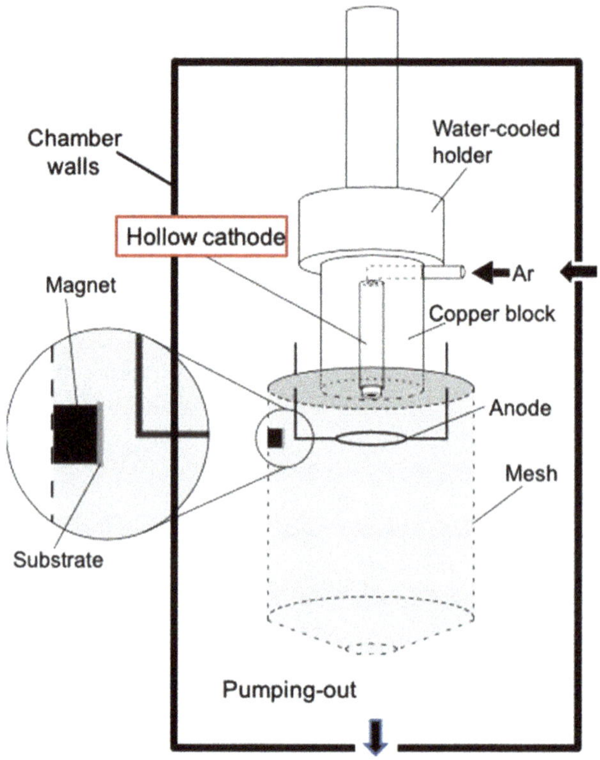

Figure 2.9. Schematic drawing of the system setup used in [60, 61]. Adapted from [60] with permission of Springer under CC BY 4.0.

chemical composition, crystallinity, morphology, size and size distribution to tailor the physical and chemical properties of the NPs. Success in obtaining tailored HNPs by a physical route encompasses various factors such as the miscibility and surface energy of the materials and the interfacial energy between the substrate and the NPs.

In fact, in the case of bimetallic NPs, for example, experimental studies and theoretical modelling have determined that the segregation, mixing and chemical ordering of metals within a bimetallic system is a result of an interplay between different intrinsic and extrinsic properties of the metals involved, as follows [62–64]:

- *Bond strength*. For two different metals A and B, if the A–B bond is stronger than the A–A and B–B bonds, this will favour the inter-mixing of elements.
- *NPs surface energies*. The material with the highest surface energy tends to occupy core positions.
- *Atomic radii*. For immiscible materials, the element with the smallest atomic radius tends to occupy the core position. However, for example, if two metals A and B are fully or partially miscible, the formation of either a solid solution or a new phase can occur. This depends on the position of the elements in the periodic table, their relative concentrations and the temperature of the alloy. In fact, for a substitutional solid solution (atoms of A replace atoms of B or

vice versa, depending on the relative concentrations of A and B) a necessary (but not sufficient) condition to accommodate an appreciable number of atoms of A in the B lattice is that the difference in atomic radii between the two elements is less than 15%. Otherwise, a new phase will form. In contrast, for an interstitial type of solid solution, the atomic radius of atom A must be substantially smaller than that of atom B.
- *Electronegativity and charge transfer.* When the elements involved in a reaction have a large electronegative difference, the probability that they will form an intermetallic compound instead of a solid solution is higher.

Other factors that can affect the formation of bimetallic NPs are the crystal structure and the valence of each element. Copper and nickel exemplify a substitutional solid solution. These two elements are completely miscible at all proportions. With regard to the above rules that govern the degree of miscibility, the atomic radii for copper and nickel are 0.128 and 0.125 nm, both have a fcc crystal structure and their electronegativities are 1.9 and 1.8, respectively. In addition, the most common valences are +1 for copper (although it can sometimes be +2) and +2 for nickel. An interstitial solid solution is formed by carbon when added to iron; the maximum concentration of carbon is approximately 2%. The radius of the carbon atom is much smaller than that of iron: 0.071 nm versus 0.124 nm [64].

Regarding examples of non-metallic HNPs prepared using the MS-IGC method, two examples of binary NPs are presented in chapter 3. The first addresses intermetallic FeAl NPs where the components are bulk-miscible (section 3.1). The second example discusses the metallic–dielectric Ag–Si binary NPs in which the materials are bulk-immiscible (section 3.2). The case of the ternary NPs Fe–Ag–Si is presented in chapter 4.

References

[1] Zheng Y, Lalander C H, Thai T, Dhuey S, Cabrini S and Bach U 2011 Gutenberg-style printing of self-assembled nanoparticle arrays: electrostatic nanoparticle immobilization and DNA-mediated transfer *Angew. Chem. Int. Edn Engl* **50** 4398–402

[2] Engstrom D S, Porter B, Pacios M and Bhaskaran H 2014 Additive nanomanufacturing – a review *J. Mater. Res.* **29** 1792–816

[3] Takashiri M, Tanaka S, Takiishi M, Kihara M, Miyazaki K and Tsukamoto H 2008 Preparation and characterization of $Bi_{0.4}Te_{3.0}Sb_{1.6}$ nanoparticles and their thin films *J. Alloys Compd.* **462** 351–5

[4] Hummers W S and Offema R E 1958 Preparation of graphitic oxide *J. Am. Chem. Soc.* **80** 1339

[5] Binns C 2009 *Handbook of Metal Physics: Metallic Nanoparticles* ed J Blackman (Amsterdam: Elsevier) ch 3

[6] Swihart M T 2003 Vapor-phase synthesis of nanoparticles *Curr. Opin. Colloid Interface Sci.* **8** 127–33

[7] Kruis F E, Fissan H and Peled A 1998 Synthesis of nanoparticles in the gas phase for electronic, optical and magnetic applications – a review *J. Aerosol Sci.* **29** 511–35

[8] Milani P and deHeer W A 1990 Improved pulsed laser vaporization for production of intense beams of neutral and ionized clusters *Rev. Sci. Instrum.* **61** 1835–8
[9] Siekmann H R, Lüder C H, Faehrmann J, Lutz H O and Meiwes-Broer K H 1991 The pulsed arc cluster ion source *Z. Phys. D: At. Mol. Clusters* **20** 417–20
[10] Haberland H, Karrais M and Mall M 1991 A new type of cluster and cluster ion source *Z. Phys. D: At. Mol. Clusters* **20** 413–5
[11] Haberland H, Karrais M, Mall M and Thurner Y 1992 Thin films from energetic cluster impact: a feasibility study *J. Vac. Sci. Technol.* A **10** 3266–71
[12] Granqvist C G and Buhrman R A 1976 Ultrafine metal particles *J. Appl. Phys.* **47** 2200–19
[13] Sattler K, Mühlbach J and Recknagel E 1980 Generation of metal clusters containing from 2 to 500 atoms *Phys. Rev. Lett.* **45** 821–4
[14] Muhlbach J, Pfau P, Recknagel E and Sattler K 1981 Cluster emission from the surfaces of Bi, Sb and Se *Surf. Sci.* **106** 18–26
[15] Pauly H 2000 *Atom, Molecule and Cluster Beams II (Springer Series on Atomic, Optical, and Plasma Physics* vol 32 (Berlin: Springer)
[16] Xirouchaki C and Palmer R E 2004 Deposition of size-selected metal clusters generated by magnetron sputtering and gas condensation: a progress review *Philos. Trans. R. Soc.* A **362** 117–24
[17] Bansmann J *et al* 2005 Magnetic and structural properties of isolated and assembled clusters *Surf. Sci. Rep.* **56** 189–275
[18] Wegner K, Piseri P, Vahedi-Tafreshi H and Milani P 2006 Cluster beam deposition: a tool for nanoscale science and technology *J. Phys. D: Appl. Phys.* **39** R439–59
[19] Binns C *et al* 2005 The behaviour of nanostructured magnetic materials produced by depositing gas-phase nanoparticles *J. Phys. D: Appl. Phys.* **38** R357–79
[20] Oxford Applied Research, http://oaresearch.co.uk/oaresearch/cluster/
[21] Private communication Mantis deposition company
[22] Yin F, Wang Z W and Palmer R E 2011 Controlled formation of mass-selected Cu–Au core–shell cluster beams *J. Am. Chem. Soc.* **133** 10325–7
[23] Xu Y H and Wang J P 2008 Direct Gas-Phase synthesis of heterostructured nanoparticles through phase separation and surface segregation *Adv. Mater.* **20** 994–9
[24] Wang J P 2008 Magnetic nanoparticles and their assembly for future magnetic media *Proc. IEEE* **96** 1847–63
[25] He S, Jing Y and Wang J P 2013 Direct synthesis of large size ferromagnetic SmCo5 nanoparticles by a gas-phase condensation method *J. Appl. Phys.* **113** 134310
[26] Krishnan G, Verheijen M A, Ten-Brink G H, Palasantas G and Kooi B J 2013 Tuning structural motifs and alloying of bulk immiscible Mo-Cu bimettalic nanoparticles by gas-phase synthesis *Nanoscale* **5** 5375–83
[27] Perez-Tijerina E, Pinilla M G, Mejia-Rosales S, Ortiz-Mendez U, Torres A and Jose-Yacaman M 2008 Highly size controlled synthesis of Au/Pd nanoparticles by inert-gas condensation *Faraday Discuss.* **138** 353–62
[28] Llamosa Perez D, Espinosa A, Martinez L, Roman E, Ballesteros C, Mayoral A, Garcia-Hernandez M and Huttel Y 2013 Thermal diffusion at nanoscale: from CoAu alloy nanoparticles to Co@Au core/shell structures *J. Phys. Chem.* C **117** 3101–8
[29] Mayoral A, Mejia-Rosales S, Mariscal M, Perez-Tijerina E and Jose-Yacaman M 2010 The Co–Au interface in bimetallic nanoparticles: a high resolution STEM study *Nanoscale* **2** 2647–51

[30] Imanaka M, Katayama T, Ohshiro Y, Watanabe S, Arai H and Nakagawa T 2004 Nanocluster ion source by plasma-gas aggregation *Rev. Sci. Instrum* **75** 1907–9

[31] Iles G N, Baker S H, Thornton S C and Binns C 2009 Enhanced capability in a gas aggregation source for magnetic nanoparticles *J. Appl. Phys.* **105** 024306

[32] Momin T and Bhowmick A 2010 A new magnetron based gas aggregation source of metal nanoclusters coupled to a double time-of-flight mass spectrometer system *Rev. Sci. Instrum.* **81** 075110

[33] Stoyanov S, Huang Y, Zhang Y, Skumryev V, Hadjipanayis G C and Weller D 2003 Fabrication of ordered FePt nanoparticles with a cluster gun *J. Appl. Phys.* **93** 7190

[34] Qiu J M, Xu Y H, Judy-Jack H and Wang J P 2005 Nanocluster deposition for high density magnetic recording tape media *J. Appl. Phys.* **97** 10P704

[35] Bai J and Wang J P 2005 High-magnetic-moment core-shell-type FeCo–Au/Ag nanoparticles *Appl. Phys. Lett.* **87** 152502

[36] Kennedy M K, Kruis F E, Fissan H, Mehta B R, Stappert S and Dumpich G 2003 Tailored nanoparticle films from monosized tin oxide nanocrystals: particle synthesis, film formation, and size-dependent gas-sensing properties *J. Appl. Phys.* **93** 551–60

[37] Fonzo F D *et al* 2000 Focused nanoparticle-beam deposition of patterned microstructures *Appl. Phys. Lett.* **77** 910

[38] Goldby I M, Issendorff B V, Kuipers L and Palmer R E 1997 Gas condensation source for production and deposition of size-selected metal clusters *Rev. Sci. Instrum.* **68** 3327

[39] Majumdar A, Köpp D, Ganeva M, Datta D, Bhattacharyya S and Hippler R 2009 Development of metal nanocluster ion source based on dc magnetron plasma sputtering at room temperature *Rev. Sci. Instrum.* **80** 095103

[40] Pellarin M, Ray C, Lermé J, Vialle J L, Broyer M and Mélinon P 2000 Gas phase study of silicon-C_{60} complexes: surface coating and polymerization *J. Chem. Phys.* **112** 8436–45

[41] Masenelli B, Tournus F, Mélinon P, Blasé X, Perez A, Pellarin M and Broyer M 2004 Nanostructured films from $(C_{60})_nSi_m$ clusters *Appl. Surf. Sci.* **226** 226–30

[42] Sumiyama K, Hihara T, Liang P D and Katoh R 2005 Structure and magnetic properties of Co/CoO and Co/Si core–shell cluster assemblies prepared via gas-phase *Sci. Technol. Adv. Mater.* **6** 18–26
Sumiyama K, Katoh R, Kadowaki S and Hihara T 2010 Fe–Si core/Si-shell clusters prepared by double glow discharge sources *J. Nanopart. Res.* **12** 2589–96

[43] Quesnel E, Pauliac-Vaujour E and Muffato V 2010 Modeling metallic nanoparticle synthesis in a magnetron-based nanocluster source by gas condensation of a sputtered vapour *J. Appl. Phys.* **107** 054309

[44] Martínez L, Díaz M, Román E, Ruano M, Llamosa D and Huttel Y 2012 Generation of nanoparticles with adjustable size and controlled stoichiometry: recent advances *Langmuir* **28** 11241–9

[45] Román-García E L, Martínez-Orellana L, Díaz-Lagos M and Huttel Y 2010 Dispositivo y procedimiento de fabricación de nanoparticulas *Spanish Patent specification* P201030059, PCT/ES2011/070032

[46] Allers T L 2013 Production of nanoparticles US2013/0270106A1 https://patents.justia.com/patent/20130270106#claims

[47] Grammatikopoulos P and Sowwan M 2017 *Gas-Phase Synthesis of Nanoparticles* ed Y Huttel (New York: Wiley-VCH Verlag GmbH & Co. KGaA) ch 14

[48] Hanuš J, Vaidulych M, Kylián O, Choukourov A, Kousal J, Khalakhan I, Cieslar M, Solař P and Biederman H 2017 Fabrication of Ni@Ti core–shell nanoparticles by modified gas aggregation source *J. Phys. D: Appl. Phys.* **50** 475307

[49] https://zerista.s3.amazonaws.com/item_files/2bee/attachments/295661/original/nanogen_brochure.pdf (Accessed November 30, 2020)

[50] Cassidy C, Singh V and Grammatikopoulos P *et al* 2013 Inoculation of silicon nanoparticles with silver atoms *Sci. Rep* **3** 3083

[51] Benelmekki M, Singh V, Baughman K W, Bohra M and Kim J-H 2021 On the potentiality of a cluster-beam-source to produce free-standing 1D and 2D FeAg nanostructures: mechanisms underlying their shape genesis *Mater. Today Chem.* **19** 100405

[52] Palmer R E, Cai R and Vernieres J 2018 Synthesis without solvents: the cluster (nanoparticle) beam route to catalysts and sensors *Acc. Chem. Res.* **51**(9) 2296–304

[53] Ellis P R, Brown C M, Bishop P T, Yin J, Cooke K, Terry W D, Liu J, Yin F and Palmer R E 2016 The cluster beam route to model catalysts and beyond *Faraday Discuss.* **188** 39–56

[54] Palmer R E, Cao L and Yin F 2016 Note: proof of principle of a new type of cluster beam source with potential for scale-up *Rev. Sci. Instrum.* **87** 046103

[55] Cai R, Jian N, Murphy S, Bauer K and Palmer R E 2017 A new method to prepare colloids of size-controlled clusters from a matrix assembly cluster source *APL Mater.* **5** 053405

[56] Cai R, Martelli F, Vernieres J, Albonetti S, Dimitratos N, Tizaoui C and Palmer R E 2020 Scale-up of cluster beam deposition to the gram scale with the matrix assembly cluster source for heterogeneous catalysis (catalytic ozonation of nitrophenol in aqueous solution) *ACS Appl. Mater. Interfaces* **12**(22) 24877–82

[57] Glotzer S C and Solomon M J 2007 Anisotropy of building blocks and their assembly into complex structures *Nat. Mater.* **6** 557–62

[58] Wagle D V and Baker G A 2015 Cold welding: a phenomenon for spontaneous self-healing and shape genesis at the nanoscale *Mater. Horiz.* **2** 157–67

[59] Zou F, Zhou H, Tan T V, Kim J, Koh K and Lee J 2015 Dual-mode SERS-fluorescence immunoassay using graphene quantum dot labeling on one-dimensional aligned magnetoplasmonic nanoparticles *ACS Appl. Mater. Interfaces* **7** 12168–75

[60] Ekeroth S, Ikeda S, Boyd R, Münger P, Shimizu T and Helmersson U 2019 Impact of nanoparticle magnetization on the 3D formation of dual-phase Ni/NiO nanoparticle-based nanotrusses *J. Nanopart. Res.* **21**(2019) 228

[61] Ekeroth S, Münger E P, Boyd R, Ekspong J, Wågberg T, Edman L, Brenning N and Helmersson U 2018 Catalytic nanotruss structures realized by magnetic self-assembly in pulsed plasma *Nano Lett.* **18**(5) 3132–7

[62] Paz-Borbon L O 2011 *Computational Analysis of Transition Metal Nanoalloys* (Berlin: Springer)

[63] Langlois C, Alloyeau D, Bouar Y L, Loiseau A, Oikawa T, Mottet C and Ricolleau C 2008 Growth and structural properties of CuAg and CoPt bimetallic nanoparticles *Faraday Discuss.* **138** 375–91

[64] Callister W D Jr 2007 *Materials Science and Engineering: An Introduction* 7th edn (Hoboken, NJ: Wiley)

IOP Publishing

Designing Hybrid Nanoparticles (Second Edition)

Maria Benelmekki

Chapter 3

Design of binary nanoparticles

In this chapter, two examples of binary nanoparticles (NPs) fabricated by magnetron-sputtering-based inert-gas-condensation (MS-IGC) method are presented. The first example addresses the synthesis and characterization of intermetallic FeAl NPs whose components are partially bulk-miscible. The advantage of working in vacuum allows a better control on the oxidation process of the NPs. The resulting NPs consisted of a nanoalloy of an iron aluminide crystalline core (DO_3 phase) encapsulated in an alumina shell that reduces inter-particle interactions and also prevents further oxidation and segregation of the FeAl core. The second example addresses the synthesis and characterization of metallic-dielectric Ag–Si binary NPs, whose materials are bulk-immiscible. In this case, an appropriate tuning of the sputtering power ratio of the Si and Ag independent targets allowed the production of NPs with distinctive multicore–shell and core-satellite morphologies. To address the recent improvement in the production yield of cluster beam sources, a third example is presented discussing the production of Au–TiO_2 and Ag–TiO_2 HNPs, which were used for catalysis purposes.

3.1 Introduction to binary NPs

The synthesis and assembly of NPs of two different materials into a binary NP can provide a general and inexpensive path to a large variety of materials (metamaterials) with precisely controlled chemical composition and tight placement of components. As discussed in section 1.2 of chapter 1, binary NPs can be combined in different configurations: (1) random or ordered alloys in a mixed state, (2) core–shell morphology (including multishell structure), (3) multicore–shell morphology, and (4) dumbbell-like and satellite morphologies in segregated or phase-separated states [1–3].

Despite the advantages that the MS-IGC method offers to produce phase-separated NPs with free-residual contamination both at the interface between the different components and on the surface, only a few studies have reported on the formation of different morphological and structural configurations for the same type of NPs with fixed chemical composition [1, 4].

Success in obtaining binary NPs with controlled structural and morphological configurations by the physical route depends strongly on the miscibility and surface energies of the elements involved in the reaction [5]. The factors affecting NP formation are discussed in more detail in section 2.3. In this chapter, three examples of binary NPs are presented. The first example addresses the synthesis and characterization of intermetallic FeAl NPs whose components are partially bulk-miscible. The resulting NPs consist of a nanoalloy of an iron aluminide crystalline core (DO_3 phase) encapsulated in an alumina shell that reduces inter-particle interactions and also prevents further oxidation and segregation of the FeAl core. The second example addresses the synthesis and characterization of metallic-dielectric Ag–Si binary NPs, whose materials are bulk-immiscible. In this case, an appropriate tuning of the sputtering power ratio of the Si and Ag independent targets allowed the production of NPs with distinctive multicore–shell and core-satellite morphologies. The NPs obtained in both cases demonstrate that the modification to the MS-IGC system made it possible to overcome the limitations that other deposition techniques face against nanoalloy oxidation and the immiscibility of materials. The third example, briefly refers the high yield production of Ag–TiO_2 and Au–TiO_2 NPs using the modified MACS system, as presented in chapter 2.

3.2 Synthesis and characterization of FeAl HNPs

3.2.1 Introduction

Current advances in nanotechnology rely increasingly on multifunctional nanostructures. Within this class of materials, soft magnetic alloys at the nanoscale level have generated intense interest because of their possible uses in many technological applications such as power transformers, inductive devices, and magnetic sensors [6, 7].

For this reason, aluminides of transition metals (e.g., nickel and iron) are promising materials due to their excellent oxidation resistance. Specifically, the FeAl system easily and quickly forms an adherent oxide layer on its surface that provides high protection to the underlying alloy when exposed to oxidizing and corrosive environments. This property, along with a soft magnetic behaviour, excellent thermal conductivity, and low mass density, makes this material a very promising candidate for many bio- and nanotechnology applications. However, traditional synthesis methods must overcome the long-standing challenge of developing precise nanoalloys with effective morphological control and stability against oxidation. In fact, when bimetallic systems are considered at the nanoscale, oxidation, phase segregation, and aggregation due to inter-particle magnetic interactions are expected, resulting in the alteration of magnetic properties and raising the question of the feasibility of soft magnetic nanoalloys [8].

Fabrication of iron aluminide alloy NPs via the MS-IGC method is described below. Note the advantage of working in vacuum, which allows a better control of the chemical purity of the NPs, specifically the oxidation process of NPs. The desired chemical composition of the NPs can be obtained by tuning the magnetron power applied on each target (Fe and Al) independently while they are co-sputtered. When the NPs are exposed to air, oxidation at high rates occurs, resulting in a pure external shell of alumina.

3.2.2 Deposition of FeAl NPs

FeAl NPs were produced using the modified MS-IGC system as described in section 2.2 of chapter 2. A supersaturated vapour of metal atoms was generated by co-sputtering iron and aluminium from two independent neighbouring targets in an argon (Ar) atmosphere. The aggregation chamber was water-cooled and evacuated to ~10^{-8} mbar prior to sputtering. Fe (99.9%) and Al (99.9995%) targets of one inch diameters were used in the DC co-sputtering process. The working pressure was maintained at 3×10^{-1} mbar in the aggregation zone. The pressure in the deposition chamber was set at 8.4×10^{-4} mbar. This differential pressure is a key factor; it determines the residence time of the NPs in the aggregation zone and therefore the crystallinity, size and shape of the NPs (effects of the residence time on the formation of NPs are discussed in section 4.5 of chapter 4). The DC powers applied to the Fe and Al targets were 11 W and 16 W, respectively. The power ratio was fixed to work in the Fe-rich part of the Fe-Al binary phase diagram in which the DO_3 and A2 phases are growth and stable at temperatures less than 500 °C. The aggregation zone length was set to 90 mm. The NPs were deposited on silicon substrates and silicon nitride TEM grids for their characterization. At the end of the deposition process, the NPs were removed from the deposition chamber and exposed to air for a few minutes before the characterization process was started.

3.2.3 Characterization methods

The size, morphology and crystal structure of these intermetallic NPs were examined using an image-corrected scanning/transmission electron microscope (S/TEM) FEI Titan 80–300 kV operated at 300 kV (see figure 1.8 of chapter 1). The chemical composition and surface oxidation of NPs were evaluated using x-ray photoelectron spectroscopy (XPS) Kratos Axis UltraDLD 39–306 equipped with a mono-AlKα source operated at 300 W (see figure 1.9 of chapter 1).

3.2.4 Morphology, structure and composition of FeAl NPs

Low magnification and high magnification TEM images (figures 3.1(a) and (b), respectively) show the core–shell morphology of the obtained NPs. Figure 3.1(f) shows the size probability distribution function (pdf) calculated over 100 NPs from TEM images. This figure reveals that approximately 42% of the NPs were 8–10 nm in size and 30% were 10–12 nm in size. The remaining 30% were 12–16 nm in size. At this point it is important to note that no size selection of NPs was applied in this work. The deposition system used in this study is equipped with a quadrupole mass filter (QMF) placed between the aggregation zone and the deposition chamber; however, for this study, it was turned off. A high-resolution transmission electron microscopy (HRTEM) image of a representative NP (figure 3.1(c)) shows its

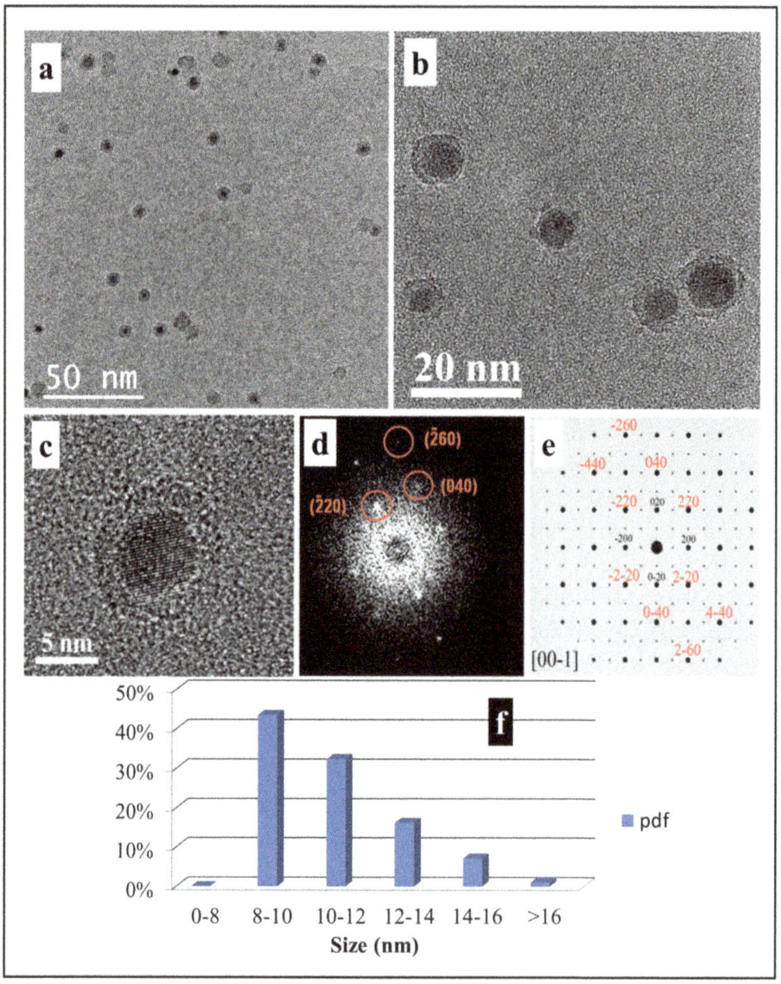

Figure 3.1. (a, b) Low- and high-magnification TEM images showing the core–shell morphology of NPs. (c) HRTEM image showing the single crystalline core encapsulated with an amorphous shell, (d) the corresponding FFT, and (e) electron diffraction pattern in the [00–1] zone axis orientation calculated by TMCrystal Maker software. (f) Size probability-distribution-function (pdf) of the NPs. (c, d, e) are reproduced from [2] with permission of AIP Publishing under CC BY 3.0.

distinctive core–shell morphology with a spherical configuration. From the HRTEM images, the shell thickness was determined to be in the range of 2–3 nm. The interplanar distance estimated from the lattice fringes was found to be 2.03 Å, which can be assigned to the Fe-rich A2, the B2, or DO3 phase. However, the high-temperature ordered B2 phase was not expected in this case due to the relatively low temperatures in an inert gas condensation technique [10]. Furthermore, the fast Fourier transform (FFT) diffractogram (figure 3.1(d)) corresponding to the HRTEM image in figure 3.1(c), combined with the simulated electron diffraction

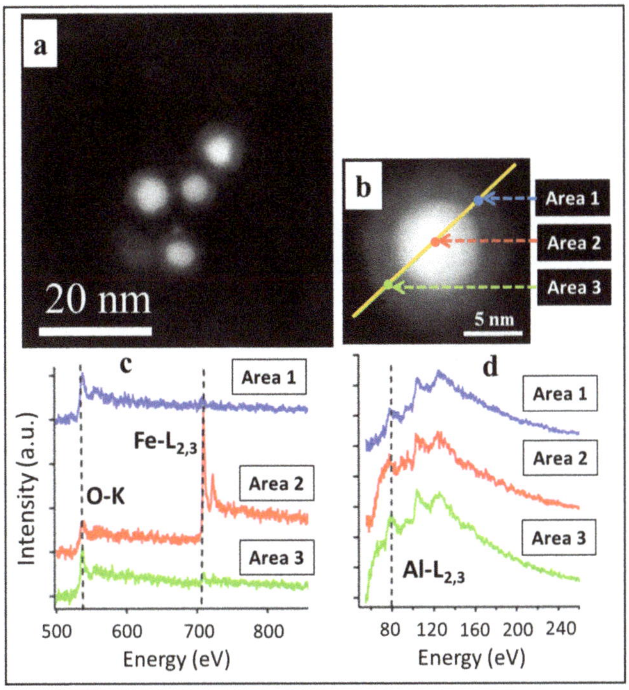

Figure 3.2. (a, b) Low- and high-magnification STEM images showing the core/shell morphology of NPs. (c, d) EELS spectra obtained from different areas of a representative NP (as indicated in (b)). EELS spectra of areas 1, 2 and 3 show the presence of Fe, Al and O within the NP. As seen in (c), area 2 shows a strong intensity of the Fe–$L_{2,3}$ edge corresponding to the position of the bright core, whilst the spectra on either side of the shell (areas 1 and 3) are dominated by the Al–$L_{2,3}$ and O–K edges. (b, c, and d) are adapted from [2] with permission of AIP Publishing under CC BY 3.0.

pattern oriented in the [00–1] zone axis (figure 3.1(e)) calculated by ™Crystal Maker software, clearly demonstrates the DO3 nature of the NP's core crystal structure.

To obtain further information on the chemical composition of these NPs, electron energy loss spectroscopy (EELs) analysis was performed (figures 3.2(c) and (d)). EEL spectra obtained from areas 1, 2 and 3 as indicated in figure 3.2(b) show the presence of Fe, Al and O within the NP. The spectra corresponding to Area-2 with brighter contrast within the NP (figure 3.2(b)) shows edges of both the Fe–$L_{2,3}$ at 707 eV (figure 3.2(c)) and Al–$L_{2,3}$ at 76 eV (figure 3.2(d)). The EEL spectra corresponding to the shell (Area-1 and Area-3) are Al- and O-rich (O–K 532 eV) areas, indicating that the shell is dominated by Al and O components.

3.2.5 An XPS study of FeAl HNPs

XPS core level spectra of Al 2p and Fe 2p are presented in figures 3.3(a) and (b), respectively. The results show that Fe and Al are present both in the metallic (73.5 eV and 706.8 eV) and the oxide (74.4 eV and 710.4 eV) states, respectively. The ratio between the peak areas of metallic Al 2p (73.5 eV) and Fe 2p (706.8 eV) is ~27%, corresponding to the DO3 phase ($Fe_{73}Al_{27}$) in the binary phase diagram of iron

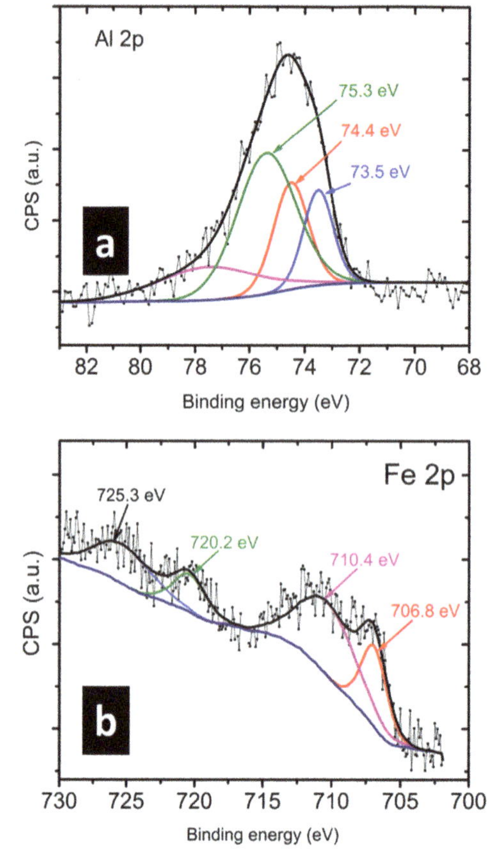

Figure 3.3. XPS spectra corresponding to (a) Al 2p and (b) Fe 2p core levels from the NPs. For the measurement, a high coverage sample of FeAl NPs deposited on Si was prepared and transferred to the XPS analysis chamber after a few minutes of exposure to atmosphere.

aluminide. Moreover, the peak corresponding to metallic Al (figure 3.3(a)) is shifted towards higher binding energy (73.4 eV instead of 72 eV), suggesting that Al atoms coordinate with Fe atoms. This matches the reported value of the Fe_3Al phase [11, 12] exactly. The peak at 75.3 eV binding energy (figure 3.3(a)) is an indication of formation of Al_2O_3 on the surface.

The same conclusion can be drawn from the O 1s peak (figure 3.4(a)) at 532.97 eV, which corresponds to the reported value for Al_2O_3 [12].

In addition, the deconvolution of the Fe 3p peak into Fe^{2+} and Fe^{3+} peaks with an atomic ratio of 1:2 (figure 3.4(b)), in combination with the Al 2p peak at 74.4 eV (figure 3.3(a)) and the O 1s peak at 531.57 eV (figure 3.4(a)), suggests the presence of spinel oxide $FeAl_2O_4$ in the inert shell after exposure to air [13, 14]. These results are in full agreement with the study reported by Graupner and co-workers [11]. In fact, when extracted from the deposition chamber and exposed to air, the NPs are oxidized at high rates, and all Al atoms available in the near-surface region will react with oxygen from air to form Al_2O_3. At ambient temperature however, the diffusion

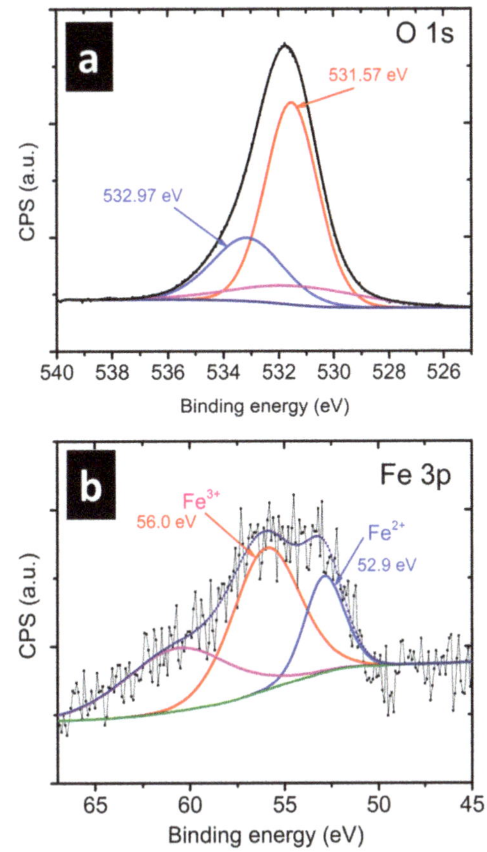

Figure 3.4. XPS spectra and curve fittings for the (a) O 1s region and (b) Fe 3p region of the sample of figure 3.3.

rate of aluminium from the core to the surface of the NPs is too low to compensate the loss of Al at the surface. The same might hold for any other mechanism involved in Al transport to the surface. Therefore, Fe-enriched regions will be exposed to the incoming oxygen, and in the absence of competition from neighbouring Al atoms, the Fe will be oxidized as well. This effect should be more pronounced when the transport of Al to the surface is slower.

The magnetic behaviour of these NPs and their stability against oxidation are presented in the original work published in reference [2]. Briefly, the NPs show soft magnetic properties with a high saturation magnetisation (170 emu g^{-1}) and low coercivity (less than 20 Oe) at room temperature. Moreover, FeAl NPs demonstrate good stability against further oxidation after 30 days of exposure to the air.

3.2.6 The formation mechanism

The size, morphology, and crystal phase of the NPs can be modulated by controlling the experimental conditions during deposition. In particular, adjusting the power of each target within the co-sputtering process allows control of the plasma densities close to the target surface; this plays an important role in nucleation, growth, and crystallisation of the NPs. In the particular experimental conditions described in this section, the low plasma density near the surface of the Fe target results in just partial nucleation and growth afterwards, when the atoms fly into the aggregation zone (see figure 2.2 of chapter 2), leading to the formation of small, amorphous Fe nanoclusters. For more details on this process, the reader can refer to section 4.7 of chapter 4 [1, 15, 16]. For Al atoms, the high concentration of electrons and ions due to the high sputtering yield induce high-density plasma near the surface of the Al target, which leads to the complete nucleation and growth of the NPs before they leave the plasma region. Subsequently, Fe and Al nanoclusters collide and coalesce with each other in the aggregation zone, forming large NPs. According to the studies reported by Yang and co-workers [17, 18], a core–shell structure is expected, with segregation of Al at the surface in most cases. However, in this case, upon exit of the plasma zone there are two competing mechanisms, the rates of which dictate the morphology of the resultant NPs. On the one hand, the important negative free enthalpy of mixing of Fe and Al [19] induces an exothermic reaction, which, combined with subsequent inter-cluster collisions, increases and maintains the high temperature of the Al nanoclusters for a period of time sufficient to create the Fe–Al crystalline alloy. On the other hand, collisions with inert gas (argon) atoms in the aggregation zone tend to cool the nanoclusters faster and consequently stabilize the single crystalline structure. When the former processes are practically complete, the latter takes over. Finally, when exposed to air, the NPs are oxidized at high rates, and all Al atoms available in the near-surface region will react with oxygen from air to form Al_2O_3 (figure 3.5). Oxidation at a low rate in the main chamber might also

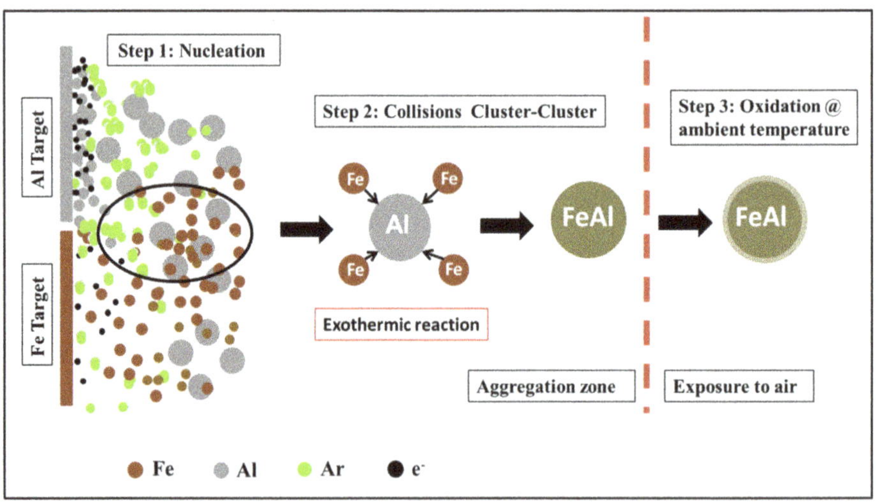

Figure 3.5. Sketch summarizing the suggested mechanism of formation of the FeAl@Al_2O_3 NPs. Step 1 shows the nucleation and growth phase in the plasma zone, near the surface of Fe and Al sputtered targets. Step 2 represents cluster–cluster collisions: the Fe and Al nanoclusters mixed in an exothermic reaction and a crystalline FeAl alloy were formed. After exposure to air, a high rate oxidation was produced, resulting in an Al_2O_3 protective shell.

occur, leading to the formation of an alumina shell; however, the formation of iron oxide (XPS spectra Fe2p and Fe3p) indicates a high oxidation rate mechanism [11].

3.3 Synthesis and characterization of AgSi NPs

3.3.1 An overview of AgSi NPs

AgSi NPs are of significant interest because the combined metal–dielectric (or metal–semiconductor) systems exhibit both photoluminescence and plasmonic properties, which make them attractive candidates for applications ranging from biosensors to informatics and storage devices. Regarding plasmonic properties, interesting studies have been reported on the tuning of the so-called local-surface-plasmon-resonance (LSPR) peak position based on Ag inter-particle coupling and the high refractive index of the dielectric matrix [20–22]. In addition, in the field of bioapplications, it has been demonstrated that the use of pure silver NPs gives rise to inconvenient liberation of Ag ions, which are toxic for the living cells [23]. However, as reported by Sotiriou and co-workers [24] for example, after coating the Ag NPs with a continuous silica shell of approximately 2 nm of thickness, the release of Ag toxic ions decreases from 12 mg l^{-1} for uncoated silver NPs to less than 2 mg l^{-1} for silica-coated NPs; the percentage of dead cells then declines from ~70% for uncoated NPs to ~10% for coated NPs.

Excluding a few exceptions, the most common reported techniques to synthesise AgSi NPs are limited to wet chemical methods. For example, Aslan and co-workers reported on controlled growth of a silica shell on Ag colloids for bioimaging applications. Their process consisted of two steps: first, silver colloids were prepared by reduction of silver nitrate by sodium citrate, and second, silica shells of various thicknesses were grown on the Ag colloids [25]. Chen and co-workers reported a modified Stober method for the synthesis of SiO_2–Ag core-satellite NPs, which can be used for plasmonic-based chemical sensing and photocatalysis applications [26]. In contrast, only a few attempts to synthesise AgSi NPs by physical methods have been reported. For example, Nomoev and Bardakhanov reported on Ag–Si core–shell NPs prepared by the electron-beam evaporation method [23]. Mertens, in his PhD thesis, reported on the synthesis of two-dimensional NP arrays embedded in an amorphous silicon (a-Si) matrix by sequential Si/Ag/Si electron-beam evaporation of Si and Ag [27].

Because the magnetron-sputtering-based gas-phase condensation method is the focal point of the present book, the immiscibility of Si and Ag in the bulk solid state illustrates the limitation of single alloy targets and then limits AgSi NP production from a single alloyed target [28, 29]. Here again, this limitation can be overcome with the modified magnetron-sputtering-based inert gas system described in figure 2.2 of chapter 2. The experimental details of the synthesis and characterization of the AgSi NPs are presented in sections 3.3.2. and 3.3.3, respectively. In section 3.3.4, a mechanism of formation of these NPs is suggested, based on the combination of the experimental conditions and the physico-chemical properties of the materials (mainly the miscibility and surface energy).

3.3.2 Deposition and characterization methods for AgSi HNPs

The AgSi NPs were synthesised by co-sputtering Si and Ag from two independent targets located side-by-side on an integrated magnetron sputtering head, as shown in figures 2.2 and 2.3 of chapter 2. The volume fraction of each component was controlled by tuning the magnetron power applied to each target independently. For all depositions, the Ar flow rate was set to 80 sccm, resulting in an aggregation zone pressure reading of 3×10^{-1} mbar. The aggregation zone length was set to 90 mm. The base pressure (7.0×10^{-8} mbar) in the main deposition chamber was increased to 3.5×10^{-3} mbar by introducing Ar flux (50 sccm). By reducing the pressure differential between the aggregation zone and the deposition chamber, the residence time of the NPs in the aggregation zone is increased, leading to the formation of larger NPs with a higher number of cores and thicker Si shells [15, 30]. The Si target sputtering power was fixed at 60 W, and the tuning of the particle composition and morphology was achieved by increasing the sputtering power supplied to the Ag target from 4.7 W to 6.7 W (~30%).

3.3.3 Morphology, structure and composition of AgSi HNPs

A high-magnification TEM image shows the multicore-shell morphology of the resulting AgSi NPs (figure 3.6(a)). The probability distribution function (pdf),

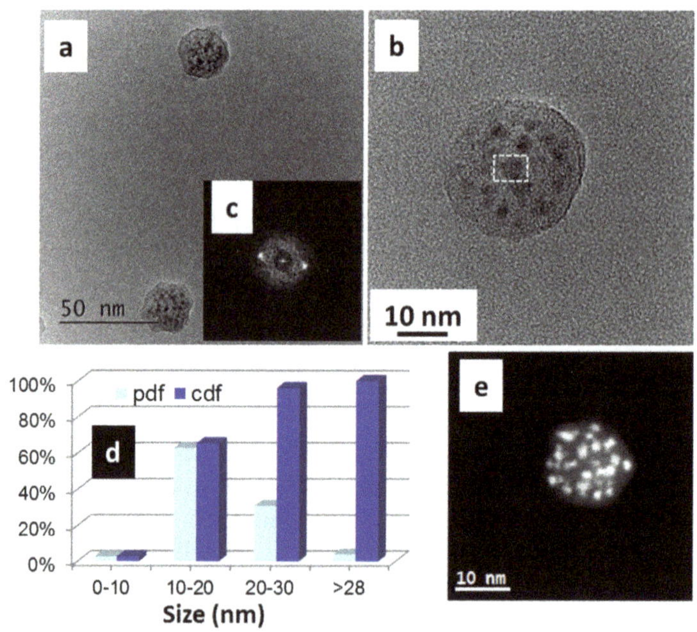

Figure 3.6. (a) High magnification TEM image showing the multicore-shell morphology of AgSi NPs. (b) HRTEM image of a representative AgSi multicore–shell NP and (c) the FFT patterns corresponding to an individual core as indicated in (b). The d-spacing measured from the pattern corresponds to the d-spacing of the {111} planes of fcc Ag. (d) Size probability distribution function (pdf) and cumulative distribution function (cdf) of the AgSi NPs. (e) STEM image showing the position of Ag cores (brighter contrast) within the Si shell.

calculated over 100 NPs, shows that approximately 60% of the NPs are 10–20 nm in size, and more than 25% are 20–30 nm in size (figure 3.6(d)). It is noteworthy that no size selection was performed during the NPs' deposition. High Resolution TEM images (figure 3.6(b)), in combination with the corresponding FFT pattern (figure 3.6(c)), show the crystalline structure of the cores. The d-spacing in the region indicated in figure 3.1(b) was determined to be 0.23 nm, corresponding to the d-spacing of the {111} planes of fcc Ag [15].

Figure 3.6(e) shows a STEM image of an individual NP in which one can observe the difference in contrast (due to the atomic numbers of Ag and Si) within the NP, indicating the positions of Ag cores (brighter contrast).

Complementary electron energy loss spectroscopic (EELS) analysis was performed to obtain further information on the chemical composition of AgSi multicore–shell NPs. From the EEL spectrum (figure 3.7(b)) corresponding to area 1 as indicated in the STEM image (figure 3.7(a)), it is possible to see the Ag–$M_{4,5}$ (401 eV) and O–K (532 eV) edges, as well as the Si–K edge at higher energy (1860 eV). The high oxygen content is attributed to the oxidation of the Si shell when the sample was exposed to the atmosphere.

In the following experiment, the power on the Ag target was increased from 4.7 W to 6.7 W while the other parameters were kept constant, as described in section 3.3.2. The resulting AgSi NPs clearly show a mixture of two types of morphologies, namely core–shell, core–satellite and core–shell–satellite structures (figure 3.8(a)). From the bright-field high-magnification TEM image (figure 3.8(c)), core–satellite NPs have a near-spherical core with a number of smaller clusters of other material deposited on their surfaces. From the high-magnification STEM image of a representative core–satellite NP (figure 3.8(b)), it is obvious that the small clusters deposited on the core are Ag-rich clusters (higher contrast = higher atomic number). The observed satellite morphology suggests that, due to the higher content of Ag in the aggregation zone, Ag clusters deposit on the surface of the NPs while in flight in the aggregation zone towards the deposition chamber.

3.3.4 The formation mechanism

This section suggests a simple mechanism of formation of AgSi based on the miscibilities and surface energies of Ag and Si. Because of its high sputtering yield (~1.20 under our experimental conditions [9]), when an Ag target is being sputtered, electrons and ions are highly concentrated in the plasma region, inducing a high-density plasma near the Ag target surface; nucleation and growth of the Ag NPs are then completed before leaving the plasma zone. In the aggregation zone, the density of inert gas is sufficient to quench their high energy. As a result, the particles retain their high temperature phase, and larger nanoclusters with a mono-crystalline structure are formed (figure 3.9 Step 1).

In contrast, silicon has a low sputtering yield of ~0.29 [9]; during Si target sputtering, the plasma density on the target surface is low, and nucleation and growth of the NPs then occur during the flight of atoms into the aggregation zone.

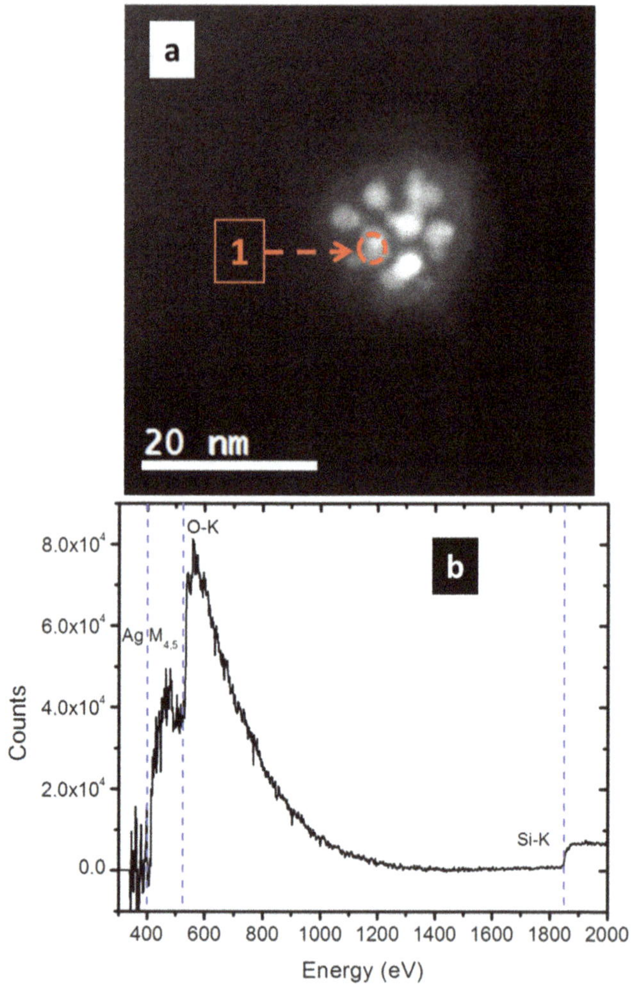

Figure 3.7. (a) STEM image of a multicore–shell AgSi NP. (b) EEL spectrum obtained from the region (1) as indicated in (a), showing the Ag $M_{4,5}$ (401 eV), O–K (532 eV), and Si–K (1860 eV) edges.

Under these conditions, Si atoms lose their energy rapidly and form small and amorphous Si nanoclusters.

Due to the low surface energy of amorphous silicon (1.05 J m^{-2}) [31], Si clusters eventually cover the surface of Ag NPs. Assisted by the lower melting temperature associated with their amorphous structure, these clusters are able to pre-melt and wet the surface of the Ag NPs, resulting in core/shell structures with shells of rather uniform thicknesses. Finally, during their flight through the aggregation zone, Ag–Si core/shell NPs collide and partially coalesce with each other, forming multicore shell morphologies. The formation of a core–satellite morphology may be attributed to the higher amount of Ag due to the increased power on the Ag target.

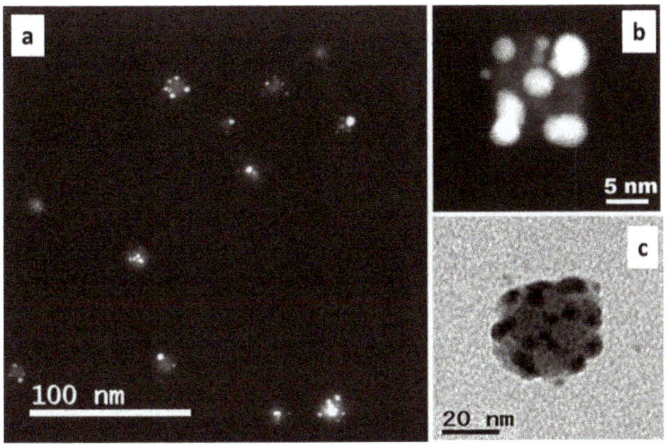

Figure 3.8. (a) STEM image showing the core–satellite, multicore–shell and core–shell–satellite morphologies of the NPs. (b) and (c) High magnification STEM and TEM images, respectively of a representative core–satellite NP.

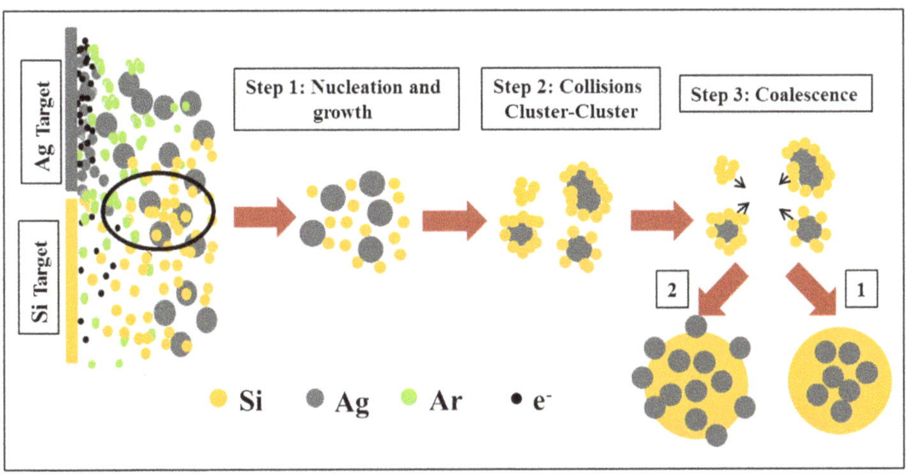

Figure 3.9. Diagram summarizing the suggested mechanism of formation of the AgSi HNPs. Step 1 shows the nucleation and growth phase in the plasma zone near the surface of the sputtered targets. Step 2 depicts cluster–cluster collisions. Si NPs then cover the surface of the Ag NPs resulting in a core/shell structure. The resulting core/shell NPs coalesce with each other in-flight and form multicore/shell HNPs (Step 3).

3.4 The production of Ag–TiO_2 and Au–TiO_2 HNPs

The Au–TiO_2 and Ag–TiO_2 NPs were prepared with the MACS shown in figure 2.8, where a photograph and schematic diagram of the system are displayed [32]. The MACS consists of a cluster deposition chamber connected to a load–lock chamber through a gate valve. In the deposition chamber, the first step consists of the generation of the matrix, by depositing Ag or Au atoms onto the Cu-cooled substrate (~20 K), while Ar gas is simultaneously introduced into the chamber.

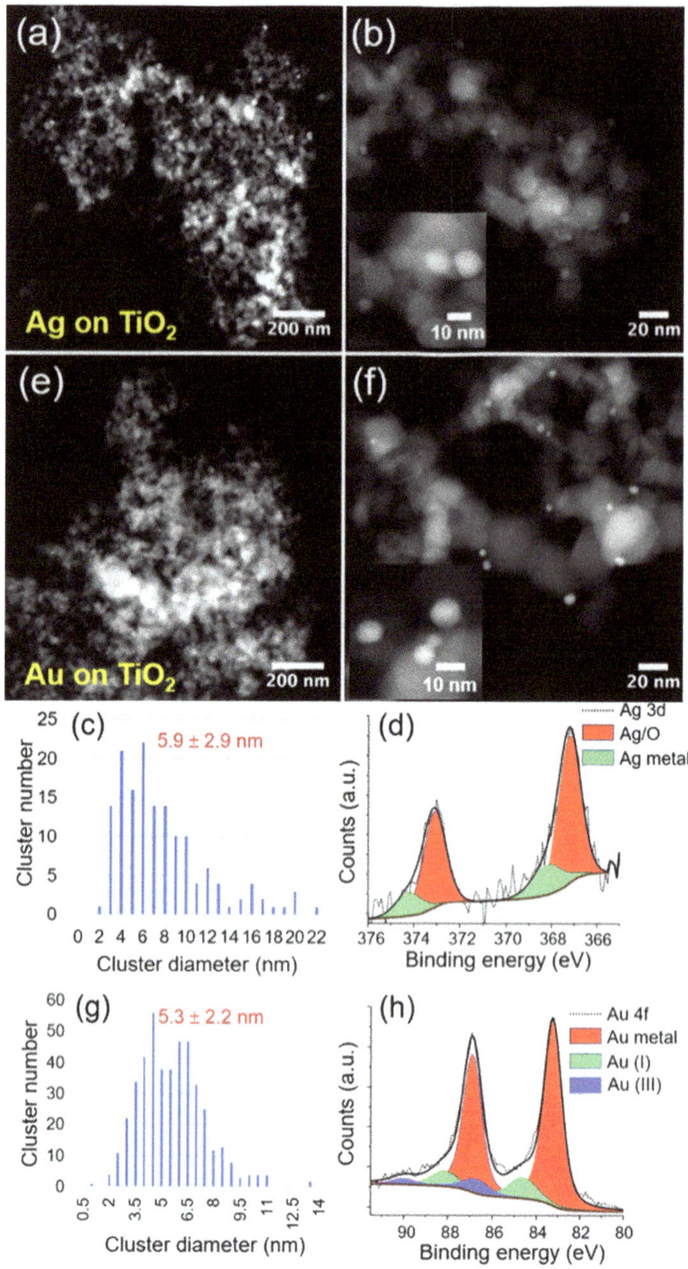

Figure 3.10. (a–b) STEM images of Ag–TiO_2 NPs. (c) The cluster diameter distributions extracted from (a–b). (d) Ag 3d core level spectrum corresponding the Ag–TiO_2 sample as shown in (a–b). (e–f) STEM images of Au–TiO_2 NPs. (g) The cluster diameter distributions extracted from (e–f). (h) Au 4f core level spectrum corresponding the Au–TiO_2 sample as shown in (e–f). Reproduced from [32] with permission of ACS under CC BY 4.0.

Subsequently, the second step starts by introducing a cup of stainless steel containing 1 g of TiO_2 powder to the deposition chamber. The Cup is placed under the matrix support to harvest the Ag or Au clusters (figure 2.8). The matrix is then sputtered with an Ar-ion beam (~2 mA) to generate a sequence of collision cascades in which the Ag or Au atoms condense together and grow to form nanoclusters. The resulting nanoclusters are sputtered out of the matrix and deposited on the TiO_2 powders in the mechanically stirred cup. In all cases, the Ag and Au nanoclusters are successfully deposited on the surface of the TiO_2, as shown in figure 3.10. Figures 3.10(d) and (h) show the Ag 3d and Au 4f high-resolution spectra for the Ag–TiO_2 and Au–TiO_2 clusters after air exposure, confirming their chemical composition.

In previous works related to older versions of MACS, such as the case presented in figure 2.6 [33]. The yield production of the clusters was demonstrated to be roughly linearly proportional to the sputtering current: ~1 cluster emitted per 100 Ar^+ ion. Based on these data, the authors assume that for this updated version of the MACS, a production of up to ~1 g of clusters (Au_{100}) per hour [32]. The obtained HNPs showed catalytic activities comparable to those of catalysts made by conventional chemical methods, indicating the near-future potential of gas-phase systems in the field of heterogeneous catalysis.

References

[1] Krishnan G, Verheijen M A, Ten-Brink G H, Palasantas G and Kooi B J 2013 Tuning structural motifs and alloying of bulk immiscible Mo–Cu bimetallic nanoparticles by gas-phase synthesis *Nanoscale* **5** 5375–83

[2] Vernieres J, Benelmekki M, Kim J H, Grammatikopoulos P, Bobo J F, Diaz R E and Sowwan M 2014 Single-step gas phase synthesis of stable iron aluminide nanoparticles with soft magnetic properties *APL Mater.* **2** 116105

[3] Singh V, Cassidy C, Grammatikopoulos P, Djurabekova F, Nordlund K and Sowwan M 2014 Heterogeneous gas-phase synthesis and molecular dynamics modeling of Janus and core-satellite Si–Ag nanoparticles *J. Phys. Chem.* C **118** 13869–75

[4] Llamosa Perez D, Espinosa A, Martinez L, Roman E, Ballesteros C, Mayoral A, Garcia-Hernandez M and Huttel Y 2013 Thermal diffusion at nanoscale: from CoAu alloy nanoparticles to Co@Au core/shell structure *J. Phys. Chem.* C **117** 3101–8

[5] Langlois C, Alloyeau D, Bouar Y L, Loiseau A, Oikawa T, Mottet C and Ricolleau C 2008 Growth and structural properties of CuAg and CoPt bimetallic nanoparticles *Faraday Discuss.* **138** 375–91

[6] Makino A, Hatanai T, Naitoh Y, Bitoh T, Inoue A and Masumoto T 1997 Applications of nanocrystalline soft magnetic materials Fe–M–B (M = Zr, Nb) alloys NANOPERM (R) *IEEE Trans. Mag* **33** 3793–98

[7] Osaka T, Takai M, Hayashi K, Ohashi K, Saito M and Yamada K 1998 A soft magnetic CoNiFe film with high saturation magnetic flus density and low coercivity *Nature* **392** 796–8

[8] Margeat O, Ciuculescu D, Lecante P, Respaud M, Amiens C and Chaudret B 2007 NiFe nanoparticles: a soft magnetic material? *Small* **3** 451–8

[9] www.iap.tuwien.ac.at/www/surface/sputteryield A simple sputter yield calculator, accessed January 2015

[10] Quesnel E, Pauliac-Vaujour E and Muffato V 2010 Modeling metallic nanoparticle synthesis in a magnetron-based nanocluster source by gas condensation of a sputtered vapour *J. Appl. Phys.* **107** 054309
[11] Graupner H, Hammer L, Heinz K and Zehner D M 1997 Oxidation of low-index FeAl surfaces *Surf. Sci.* **380** 335–51
[12] Moulder J F, Stickle W F, Sobol P E and Bomben K D 1992 *Handbook of X-ray Photoelectron Spectroscopy* ed J Chastain (Waltham, MA: Perkin Elmer Corporation)
[13] Yamashita T and Hayes P 2008 Analysis of XPS spectra of Fe^{2+} and Fe^{3+} ions in oxide materials *Appl. Surf. Sci.* **254** 2441–9
[14] Rodriguez G A C, Guillen G G, Palma M I M, Roy T K D, Hernandez A M G, Krishnan B and Shaji S 2014 Synthesis and characterization of Hercynite nanoparticles by pulsed laser ablation in liquid technique *Int. J. Appl. Ceram. Tec.*
[15] Benelmekki M, Vernieres J, Kim J H, Diaz R E, Grammatikopoulos P and Sowwan M 2015 On the formation of ternary metallic-dielectric multicore-shell nanoparticles by inert-gas condensation method *Mater. Chem. Phys.* **151** 275–81
[16] Xu Y H and Wang J P 2008 Direct gas-phase synthesis of heterostructured nanoparticles through phase separation and surface segregation *Adv. Mater.* **20** 994–9
[17] Yang J, Hu W, Tang J and Dai X 2013 The formation of FecoreAlshell and FeshellAlcore nanoparticles, a molecular dynamics simulation *Comp. Mater. Sci.* **74** 160–4
[18] Yang J, Hu W and Tang J 2014 Effect of incident energy on the configuration of Fe–Al nanoparticles, a molecular dynamics simulation of impact deposition *RSC Adv.* **4** 2155–60
[19] Breuer J 2001 Statistical thermodynamics of ordered intermetallic compounds containing point defects *PhD Thesis* University of Stuttgart, Germany
[20] Alqudami A and Annapoorni S 2007 Fluorescence from metallic silver and iron nanoparticles prepared by exploding wire technique *Plasmonics* **2** 5–13
[21] Zhai Y, Han L, Wang P, Li G, Ren W, Liu L, Wang E and Dong S 2011 Superparamagnetic plasmonic nanohybrids: shape-controlled synthesis, TEM-induced structure evolution, and efficient sunlight-driven inactivation of bacteria *ACS Nano* **5** 8562–70
[22] Cade N I, Ritman-Meer T, Kwakwa K A and Richards D 2009 The plasmonic engineering of metal nanoparticles for enhanced fluorescence and Raman scattering *Nanotechnology* **20** 285201
[23] Nomoev A V and Bardakhanov S P 2012 Synthesis and structure of Ag–Si nanoparticles obtained by the electron-beam evaporation/condensation method *Tech. Phys. Lett.* **38** 375–8
[24] Sotiriou G A, Hirt A M, Lozach P Y, Teleki A, Krumeich F and Pratsinis S E 2011 Hybrid, silica-coated, Janus-like plasmonic-magnetic nanoparticles *Chem. Mater.* **23** 1985–92
[25] Aslan K, Wu M, Lakowicz J R and Geddes C D 2007 Fluorescent core–shell Ag@SiO2 nanocomposites for metal-enhanced fluorescence and single nanoparticle sensing platforms *J. Am. Chem. Soc.* **129** 1524–5
[26] Chen K H, Pu Y C, Chang K D, Liang Y F, Liu C M, Yeh J W, Shih H C and Hsu Y J 2012 Ag-nanoparticle-decorated SiO_2 nanospheres exhibiting remarkable plasmon-mediated photocatalytic properties *J. Phys. Chem.* C **116** 19039–45
[27] Mertens H 2007 Controlling plasmon-enhanced luminescence PhD Thesis Institute for Atomic and Molecular Physics, Amsterdam
[28] Olesinski R W, Gokhale A B and Abbaschlan G J 1989 The Ag–Si (Silver–Silicon) system *Bull. Alloy Phase Diagr* **10** 635–40

[29] Bokhonov B and Korchagin M 2002 In-situ investigation of the formation of eutectic alloys in the systems silicon–silver and silicon–copper *J. Alloys Compd.* **335** 149–56

[30] Benelmekki M, Bohra M, Kim J H, Diaz R E, Vernieres J, Grammatikopoulos P and Sowwan M 2014 Facile single-step synthesis of ternary multicore magneto-plasmonic nanoparticles *Nanoscale* **6** 3532–5

[31] Hara S, Izumi S, kumagai T and Sakai S 2005 Surface energy, stress and structure of well-relaxed amorphous silicon: a combination approach of *ab initio* and classical molecular dynamics *Surf. Sci.* **585** 17–24

[32] Cai R, Martelli F, Vernieres J, Albonetti S, Dimitratos N, Tizaoui C and Palmer R E 2020 Scale-up of cluster beam deposition to the gram scale with the matrix assembly cluster source for heterogeneous catalysis (catalytic ozonation of nitrophenol in aqueous solution) *ACS Appl. Mater. Interfaces* **12** 24877–82

[33] Ellis P R, Brown C M, Bishop P T, Yin J, Cooke K, Terry W D, Liu J, Yin F and Palmer R E 2016 The cluster beam route to model catalysts and beyond *Faraday Discuss.* **188** 39–56

Chapter 4

Design of ternary magneto-plasmonic nanoparticles

In this chapter, magneto-plasmonic nanoparticles (NPs) composed of multiple FeAg cores encapsulated by a Si shell are prepared by magnetron-sputtering-based inert-gas-condensation (MS-IGC) method. Independent targets of Fe, Ag, and Si were co-sputtered under controlled conditions allowing the tuning of the size, chemical composition, and morphology of the NPs. In comparison to the widely reported binary magneto-plasmonic NPs, the Si shell maintains a fixed spatial configuration between the multiple cores and prevents FeAg cores from oxidising. When these NPs are exposed to the atmosphere, the Si surface is oxidized to SiO_x, providing an easy path to functionalize their surface for a wide range of bioapplications. This methodology can be extended to materials others than those addressed here. In addition, the chapter presents new examples related to recent development of magnetic and plasmonic trimetallic NPs (Au–Pt–Pd and FeRh@Au) produced by gas phase method.

4.1 Introduction to magneto-plasmonic NPs

Hybrid NPs (HNPs) that contain both magnetic and optically active components have emerged as attractive candidates for advanced biomedical applications, such as multimodal bio-imaging, targeted drug delivery and magnetic hyperthermia. In particular, dumbbell-like NPs composed of magnetite (Fe_3O_4) and Au or Ag, are attractive because of their biocompatibility, plasmonic activity and magnetic properties [1–7]. A broad range of techniques in chemistry and physics has been explored for the preparation of such multicomponent NPs. Further details of the synthesis, properties and applications of dumbbell-like NPs can be found in several review articles [1, 8, 9].

In the last decade, physical vapour deposition techniques have been developed for production of hybrid inorganic NPs. However, only a few attempts have been made

doi:10.1088/978-0-7503-3587-4ch4

to produce heterogeneous, dumbbell-shaped multicomponents [10, 11] other than the widely reported core–shell nanostructures [12, 13]. In the context of this chapter, reports of the preparation of Fe-rich-core–Si-rich-shell NPs using double-glow discharge sources [14, 15] are worth noting. The behaviour of Ag–Fe magneto-plasmonic NPs (MPNPs) prepared by magnetron sputtering with subsequent in-flight annealing was reported in [16] and demonstrated the instability of these NPs under atmospheric conditions.

4.2 The deposition of FeAg@Si MPNPs

NPs were synthesized using the MS-IGC method as described in chapter 2. Three independent targets of Fe, Ag and Si were co-sputtered under controlled conditions (figures 2.2 and 2.3). One-inch diameter Si (99.999%), Ag (99.99%) and Fe (99.9%) targets were used as the ion sources for NP formation. The DC powers applied to the Si, Fe and Ag targets were 64 W, 10 W and 4.8 W, respectively. Ar flux in the aggregation zone was set at 80 sccm resulting in a working pressure of 3×10^{-1} mbar in the aggregation zone. The pressure at the deposition chamber was set at 7.5×10^{-4} mbar. The NPs were deposited on different substrates for TEM images: Si, Cu, graphite and carbon coated Cu grids. The working distance, from the target surfaces to the exit slit (figure 2.2(a)), was set at ~90 mm.

4.3 Characterization methods

TEM analysis was performed using a Cs-corrected-FEI-Titan 80–300 kV microscope operating at 300 kV (see figure 1.9). SEM images were collected using an FEI Quanta FEG 250 system. For XPS analysis, a Kratos Axis Ultra^{DLD} 39–306 electron spectrometer equipped with a monochromatic AlKα source operated at 300 W was used (figure 1.10).

4.4 The morphology, structure and composition of FeAg@Si MPNPs

The size PDF and CDF of the resulting NPs were calculated from TEM images. As shown in figure 4.1(b), approximately 65% of the resulting MPNPs (hereafter MPNPs-1) were 10–20 nm in size and 20% were 20–30 nm in size. This size distribution can be correlated with the number of cores per NP, as reported in [17]. Figures 4.1(c)–(e) display high magnification TEM images of individual MPNPs showing multicore–shell morphology and irregular shapes. STEM images (figures 4.1(f)–(h)) show the differences in contrast (due to the atomic numbers of the constituents) within MPNPs-1, indicating the positions of Ag NPs (brighter contrast). HRTEM images of a representative four-core NP (figures 4.2(a-1) and (a-2)) show crystalline structure differences within the individual dumbbell-like shaped cores embedded in an amorphous shell. A STEM image of the same NP (figure 4.2(b)) shows the difference in contrast within the cores, indicating the positions of Ag NPs (brighter contrast).

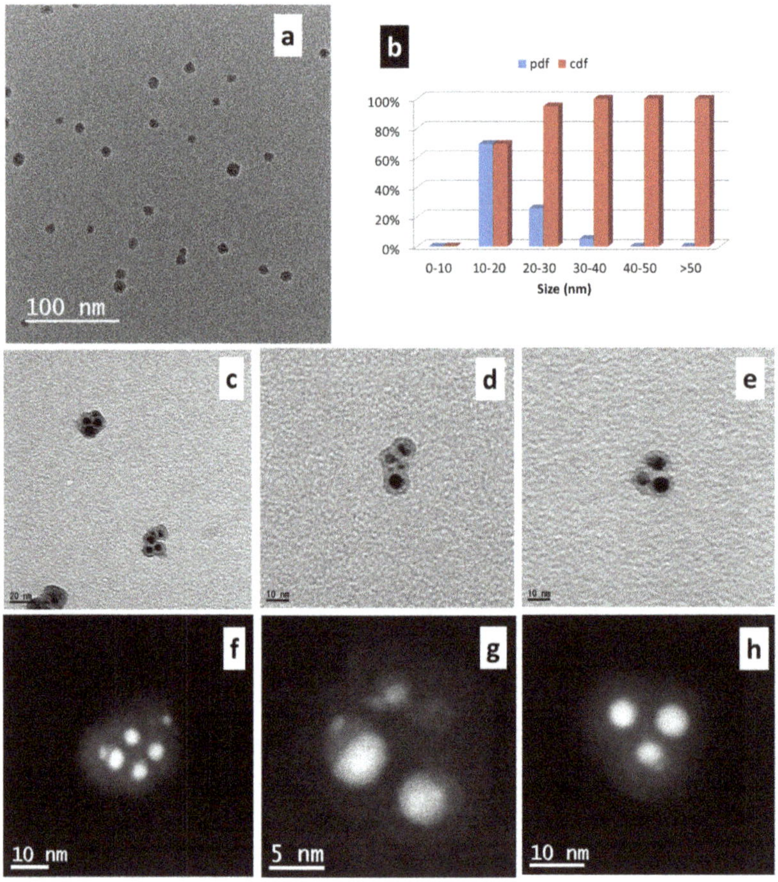

Figure 4.1. (a) Low magnification TEM image, (b) size probability distribution function (pdf) and cumulative distribution function (cdf) of MPNPs-1, calculated from TEM images. (c–e) Bright-field TEM images of individual FeAg@Si NPs showing the multicore/shell morphology and the irregular shapes of the NPs. (f–h) STEM images of FeAg@Si NPs showing the dumbbell-shaped cores. Brightness variations in (f–h) images indicate three different materials.

For further information on the chemical composition of the cores, EELS analysis was performed on a two-core NP, as shown in figure 4.3. Areas with brighter contrast in the dumbbell-like NP are Ag-rich. In fact, from the EELS spectrum corresponding to area d (figure 4.3(d)), it is possible to distinguish both the Ag $M_{4,5}$ (401 eV) and O–K (532 eV) edges, as well as a small Fe $L_{2,3}$ (710 eV) edge that can result from a small amount of Fe trapped in the Si shell. However, in the EELS spectrum corresponding to area c (figure 4.3(c)), only the Fe $L_{2,3}$ edge and O–K edge are present (the oxygen arises from oxidation of the Si shell when exposed to air). In both cases, the EELS spectra show a Si edge at higher energy (1860 eV). From the HRTEM image (figure 4.3(b)) and associated FFTs (not shown), the Ag-rich region reveals the {111} Ag planes with an interplanar distance of 0.238 nm, while the

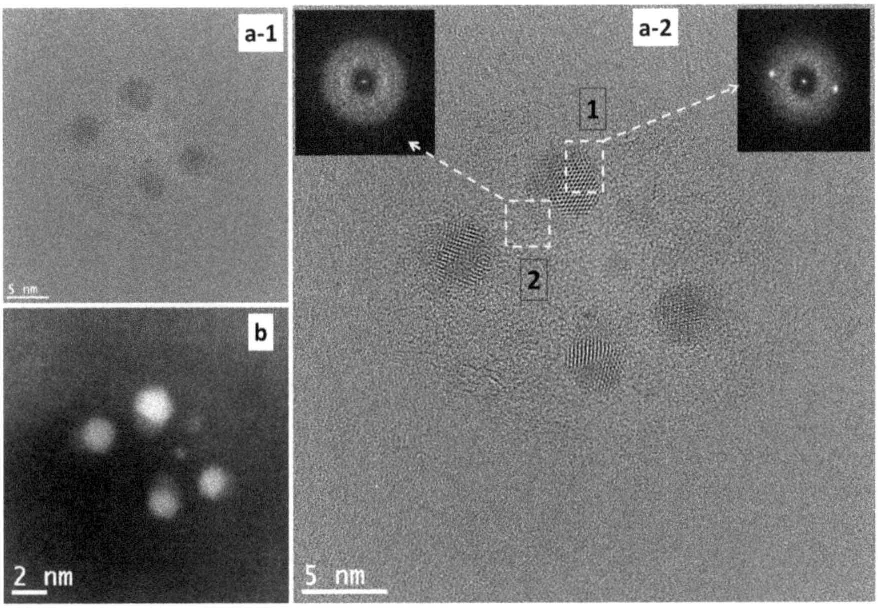

Figure 4.2. (a-1) and (a-2) High resolution TEM images of the same four-core NP showing the differences in contrast within the dumbbell-like core and the Si amorphous shell. Insets in figure (a-2) show the FFT patterns corresponding to regions 1 and 2. The d-spacing measured from pattern 1 corresponds to the d-spacing of the {111} planes of fcc Ag. FFT pattern 2 demonstrates the amorphous structure of the Fe. (b) STEM image of the same NP showing the brightness variation within the cores indicating two different materials.

Fe-rich and Si regions show an amorphous structure. Thus, this combination of techniques, i.e., bright field imaging, STEM imaging and EELS analysis, confirm the core–shell structure of the nanoclusters with a core of Fe–Ag dumbbell-like NPs covered with a Si shell.

Another confirmation of the dumbbell-like structure of the cores can be achieved by scanning across one core, as shown in figure 4.4. The line scan through the core, in the direction of the arrow as indicated in figure 4.4(a), shows an increasing Ag signal that reaches a peak at approximately 4 nm and then decreases to a minimum at approximately 7.5 nm. For the Fe signal, a defined minimum at ~4 nm was observed, followed by a distinct maximum at ~7.5 nm.

4.5 The morphology and size tuning of MPNPs

4.5.1 The effects of the pressure differential between the aggregation zone and the deposition chamber

Increasing the pressure in the deposition chamber from 7.5×10^{-4} mbar to 3.9×10^{-3} resulted in the formation of MPNPs termed MPNPs-2 (figure 4.5(a)). In comparison with MPNPs-1, only 50% of MPNPs-2 are below 30 nm in size, 20% are 30–50 nm and 30% exceeded 50 nm (figure 4.5(b)). These results demonstrate that by reducing the pressure differential between the aggregation zone and the deposition chamber, the

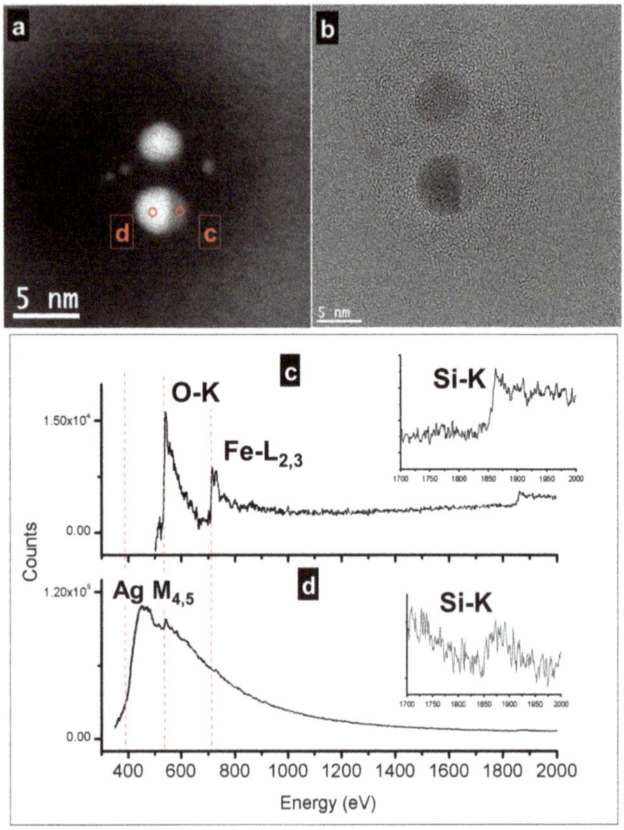

Figure 4.3. (a) STEM and (b) HRTEM images of the same NP. (c, d) EEL spectra showing that the cores are composed of Fe and Ag domains in a dumbbell-like shape and that the shell is composed of Si [17]. Reproduced by permission of The Royal Society of Chemistry.

residence time of the NPs in the aggregation zone increases and the coalescence process is enhanced, leading to the formation of larger NPs with a higher number of cores and thicker Si shells (figures 4.5(c)–(e)) [18]. EELS profiles were intended to determine the position of each component within the NPs; however, even for the smallest MPNPs-2, due to the thickness of the Si shell and the larger sizes of the NPs, it was difficult to determine whether the cores had a core–shell structure or a dumbbell-like shape (figure 4.6).

4.5.2 The effects of discharge powers on Ag target

In this experiment, the power on the Ag target was increased from 4.8 W to 6.7 W while the other parameters were kept constant, as described in section 4.2. The resulting MPNPs (hereafter MPNPs-3) clearly have a mixture of two types of structures, namely core–shell and core–satellite structures (figures 4.7(a) and (b)). The core–shell NPs have irregular shapes with well-defined cores of one material

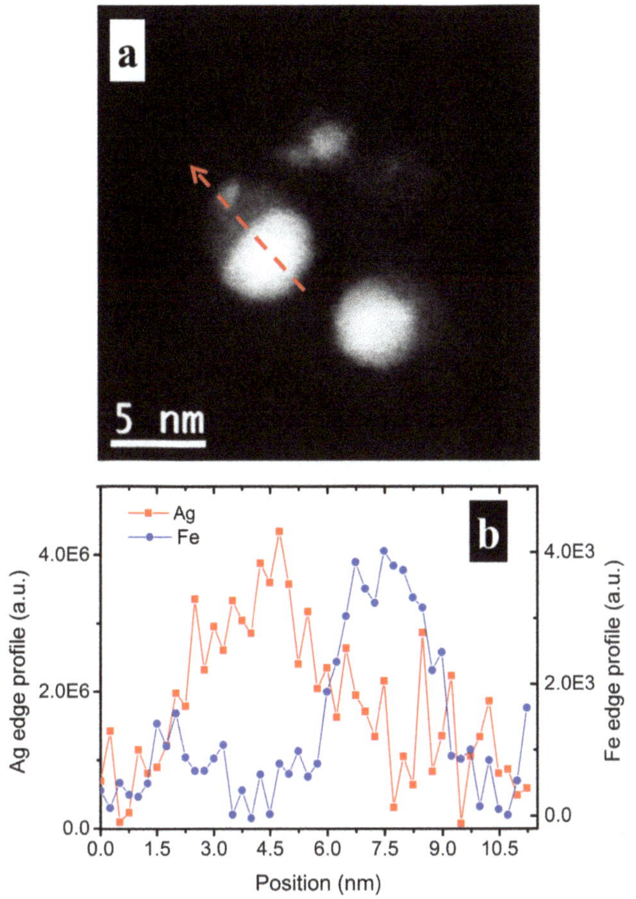

Figure 4.4. (a) STEM image and (b) its corresponding EELS line scan showing the dumbbell-like structure of the Ag–Fe core. Red arrow in STEM image represents the path where the beam was scanned.

encapsulated by another material. The core–satellite NPs have a roughly spherical core with a number of smaller clusters of other material deposited on their surfaces. The observed satellite morphology suggests that, because of the higher content of Ag in the aggregation zone, Ag clusters deposit on the surface of the formed NPs while in flight in the aggregation zone. Size distributions were obtained from TEM images. 80% of the NPs were 10–20 nm in size, clearly smaller than MPNPs-1 (figure 4.7(e)).

EELS line scans were performed across the particle in the direction of the arrow, as shown in figure 4.8. The Ag and Fe intensity profiles both show maxima at ~4.5 nm and ~9 nm positions, indicating a core–shell structure in both core NPs. The narrow shoulders of the Fe signal, appearing at ~7.5 and ~10.5 nm, indicate depletion of Fe during the growth process. These results suggest that by increasing the power on the Ag target, the Ag NPs formed have sufficient energy to mix with iron and segregate to form Fe–Ag core–shell NPs [11].

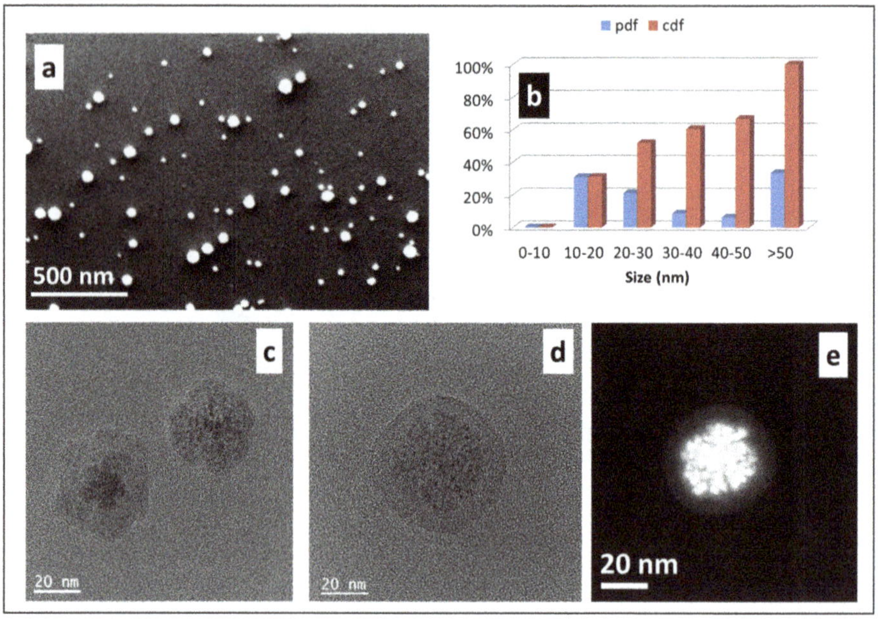

Figure 4.5. (a) SEM image, (b) size probability distribution function (pdf), and cumulative distribution function (cdf) of MPNPs-2, calculated from the SEM image. (c, d) TEM images of different NPs, (d) STEM image showing the multicore–shell morphology of resulting NPs.

4.6 The oxidation state of MPNPs

XPS measurements were performed after exposure of the samples to the atmosphere. Fe $2p_{3/2}$ fitting peaks suggest that the iron cores in MPNPs-1 were dominated by the Fe^{3+} state (711.12 eV) (figure 4.9(b)). A first approximation of the relative contents of Fe, Ag and Si was calculated from Ag 3d, Fe 2p and Si 2p XPS core level peak areas (figure 4.9). The weight percentages (wt%) obtained of Fe, Ag and Si were ~33.82%, ~48.66%, ~17.51%, respectively. However, it is important to note that because of the surface oxidation effects, this calculation is not accurate, and thus is simply a rough approximation of the composition of the MPNPS-1 sample.

In contrast, in MPNPs-2 and MPNPs-3 the iron was present in both metallic (707.12 eV) and oxidized (711.33 eV) states [19, 20] (figures 4.10(d) and (f)). For Si 2p fitting peaks (figures 4.10(a), (c) and (e)), the atomic ratios between the Si2p@Si (99.11 eV) and Si2p@SiO_x (102.10 eV) were ~20% for MPNPs-1 and ~43% for MPNPs-2. In the case of the MPNPs-3, the Si shell was nearly fully oxidized to SiO_x. These results suggest that in the case of MPNPs-2, because the Si shell was thicker than that of MPNPs-1 and MPNPs-3, a gradual oxidation occurred from the surface of the Si shell to the interface between the Si and the FeAg cores, providing better protection for the Fe cores against oxidation. In the case of MPNPs-3, despite the full oxidation of the Si shell, Fe was present in both its metallic and oxidized states. This behaviour was in agreement with EELs profiles, confirming that the iron was protected by the Ag shell (figure 4.8).

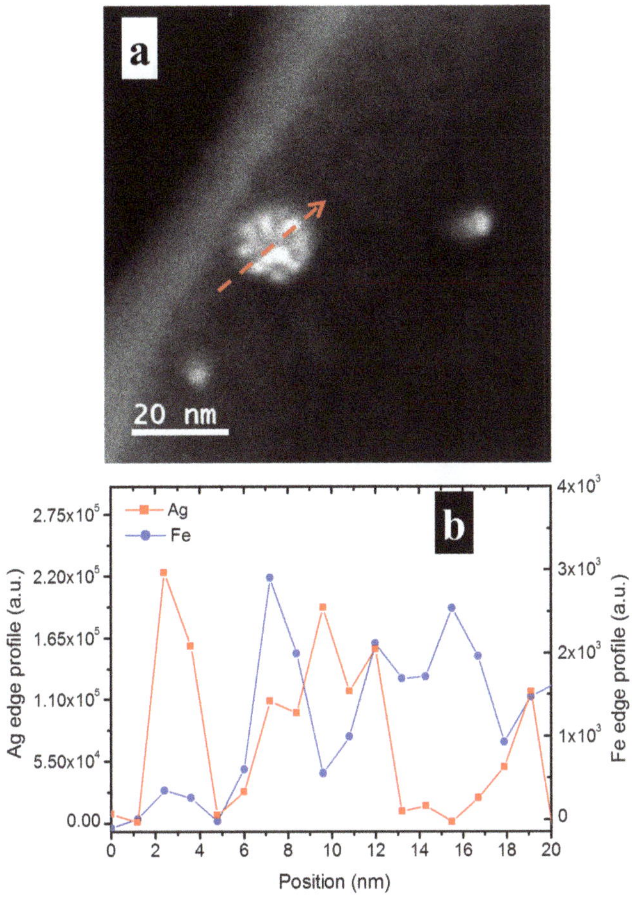

Figure 4.6. (a) STEM image and (b) its corresponding EELS line scan of a NP selected from MPNPs-2 sample. From these profiles, it is difficult to confirm the morphology of the cores. Red arrow in STEM image represents the path of the scanning beam.

Regarding magnetic properties, the MPNPs exhibit typical a ferromagnetic behaviour with a coercivity of ~102 Oe at 300 K. These coercivities in all samples can be attributed to the anisotropy at the FeAg interface. In addition, the UV–Visible spectra corresponding to MPNPs show an enhanced, red shifted, light absorption band due to the strong near-field coupling between the Ag cores and the Si shell. For more details on magnetic and optical properties, the reader can consult references [17, 18, 25].

4.7 The formation mechanism

Based on the fundamentals of magnetron sputtering, when the targets of Fe, Ag and Si are simultaneously sputtered, the plasma density plays a major role in the nucleation, growth and crystallization of the NPs. As previously explained (section 3.3.4), under the current experimental conditions, large nanoclusters of Ag with a

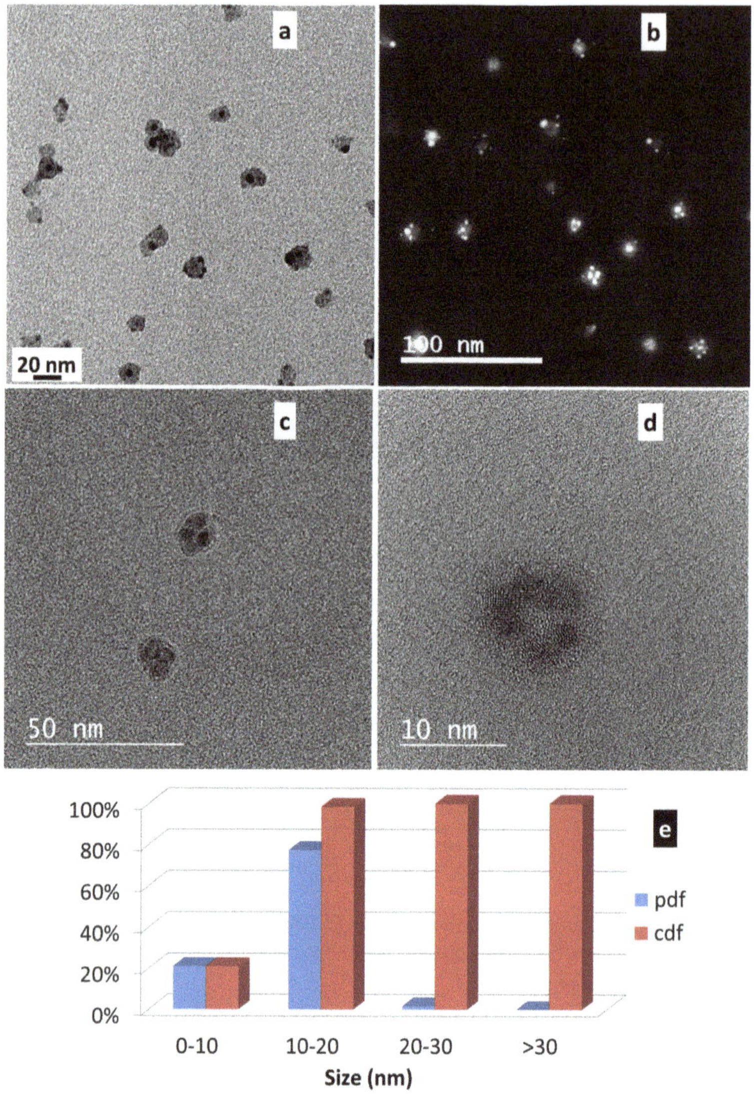

Figure 4.7. Low magnification TEM (a) and STEM (b) images of MPNPs-3 showing both core/shell and satellite morphologies. (e) The size and cumulative size probability distribution functions (pdf and cdf) of MPNPs-3 calculated from TEM images. (c) High magnification and (d) high resolution TEM images, respectively, showing the proximity of cores inside a Si shell.

mono-crystalline structure are formed. This is a result of the combination of two mechanisms: (1) the high-density plasma near the surface of the Ag target and (2) the sufficiency of the inert gas to quench the high energy of the formed Ag NPs towards the aggregation zone (figure 4.11, step 1).

In contrast, for Fe NPs, the estimated sputtering yield is ~0.45 [23]. Moreover, the Fe target provides a path for the magnetic flux generated by the permanent magnet of the cathode (figure 2.4). Flux passes through the Fe target (figure 2.4(c))

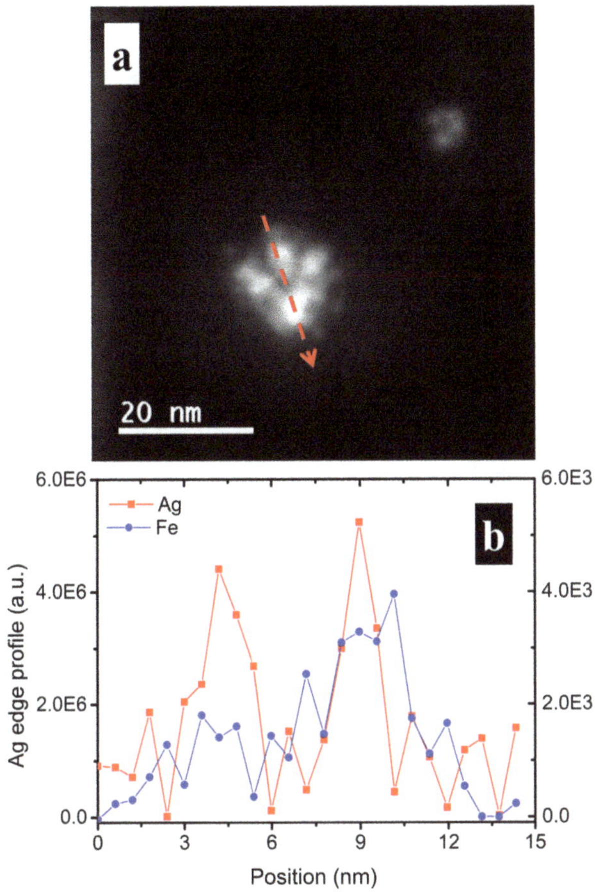

Figure 4.8. (a) STEM image and (b) its corresponding EELS line scan showing the core–shell morphology of the Ag-Fe core. Red arrow in STEM image represents the path where the beam was scanned.

and only a small portion escapes and reaches the surface, contributing to the generation of the magnetron plasma. Since the density and the distribution of the plasma are strongly correlated with the magnetic field strength and magnetic flux distribution [21, 22], during Fe sputtering only a low concentration of electrons near the surface is achieved. As a result, the ionization rate is lower than that of the Ag target, leading to a lower density of plasma. In this case, nucleation and growth of NPs occur during the flight of atoms into the aggregation zone. There, Fe atoms lose their energy rapidly and form small, amorphous Fe nanoclusters. The sputtering yield of Si is determined to be ~0.29 [23], inducing a low plasma density near the surface of the Si target. These experimental conditions lead to the formation of small and amorphous Si NPs.

Because the surfaces of the formed NPs are not oxidized, and the fraction of surface atoms per unit volume is very large, in the aggregation zone nanoclusters collide and coalesce by diffusion of the constituent atoms at their contact interfaces, creating larger NPs. Due to the large, positive free energy of Fe and

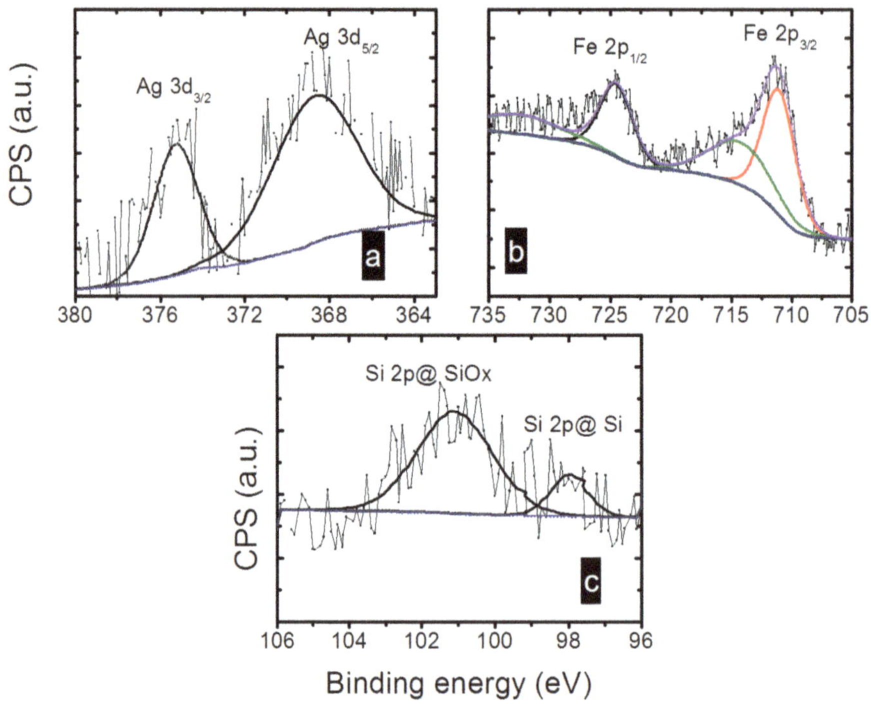

Figure 4.9. XPS spectra corresponding to (a) Ag 3d, (b) Fe 2p and (c) Si 2p core levels of MPNPs-1 sample.

Ag mixing, segregation of the elements forming core–shell or dumbbell-like structures would be expected [11, 16]. However, from our TEM study, the almost exclusive formation of dumbbell-like structures was observed in MPNPs-1. A plausible explanation for this behaviour is that when Fe and Ag NPs collide in the aggregation zone, their energy is not sufficient to induce complete coalescence, so that only dumbbell-like structures can be formed. In the case of MPNPs-3, due to the higher current discharge applied to the Ag target, the pre-formed Ag NPs seem to retain enough energy to collide and mix with the Fe NPs, and then segregate to form an Fe–Ag core–shell structure (figure 4.11, step 2). Furthermore, due to the low surface energy of amorphous silicon (1.05 J m^{-2}) [24], Si clusters eventually cover the surface of the FeAg NPs. In addition, the lower melting temperature associated with the Si amorphous structure allows these clusters to pre-melt and wet the surface of the FeAg NPs, resulting in core–shell structures with shells of quite uniform thicknesses. Finally, during their flight through the aggregation zone, the FeAg–Si core–shell NPs collide and partially coalesce with each other, forming multicore MPNPs.

When the pressure differential between the aggregation zone and the main chamber is reduced, the residence time of MPNPs in the aggregation zone is increased, increasing the probability of coalescence, resulting in larger NPs with a higher number of cores and thicker Si shells.

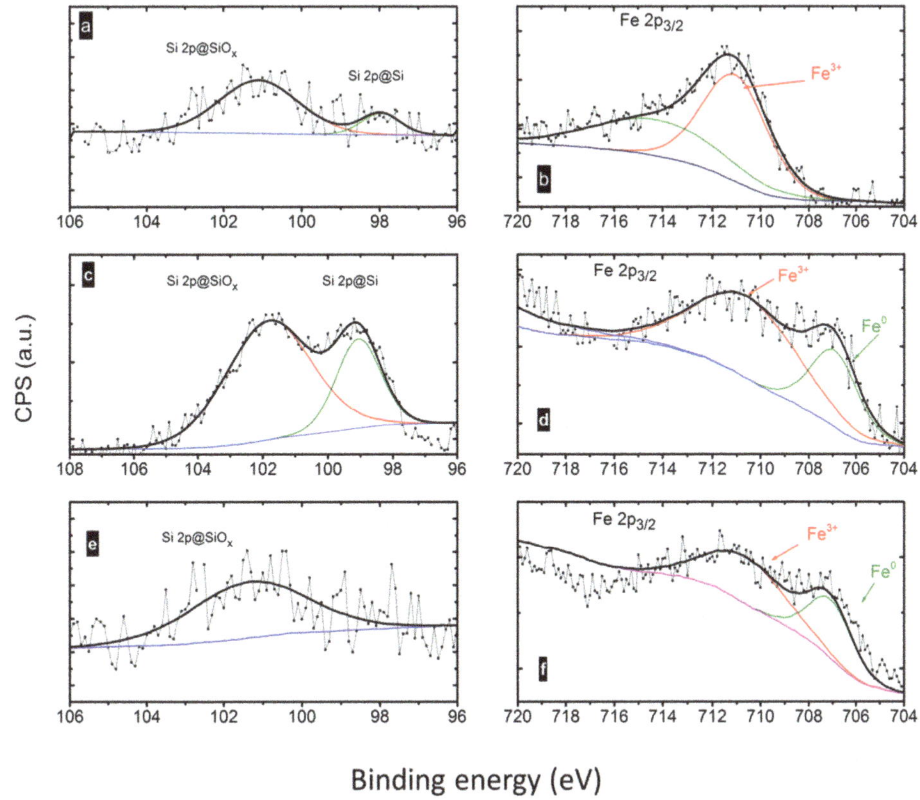

Figure 4.10. XPS spectra corresponding to Si 2p and Fe 2p core levels from MPNPs-1(a, b), MPNPs-2 (c, d), and MPNPs-3 (e, f).

4.8 Trimetallic magnetic and plasmonic HNPs

4.8.1 Introduction

Multicomponent NPs are attractive for a wide range of applications catalysis being one them. Particularly, multimetallic NPs such as Pt-based NPs have been extensively utilized in the chemical industry, due to the remarkable catalytic activity of Pt [26, 27]. Within this context, it was suggested that the enhancement of Pt catalytic activity can be achieved by the addition of other metallic components such as Au or Pd, resulting in multielement HNPs, and thus various studies focused on the Au–Pt–Pd system developed by chemical route [28–31]. Although the control of the chemical composition of multielement HNPs was achieved through efficient synthesis techniques, the spatial control of the position of each element remains challenging [32, 33]. On the other hand, intermetallic NPs offer a better resistance to oxidation and display enhanced properties. However, their synthesis appears even more challenging at the nanoscale as the chemically ordered intermetallic alloys usually form at high temperatures, which are difficult to reach in many synthesis routes [34–36].

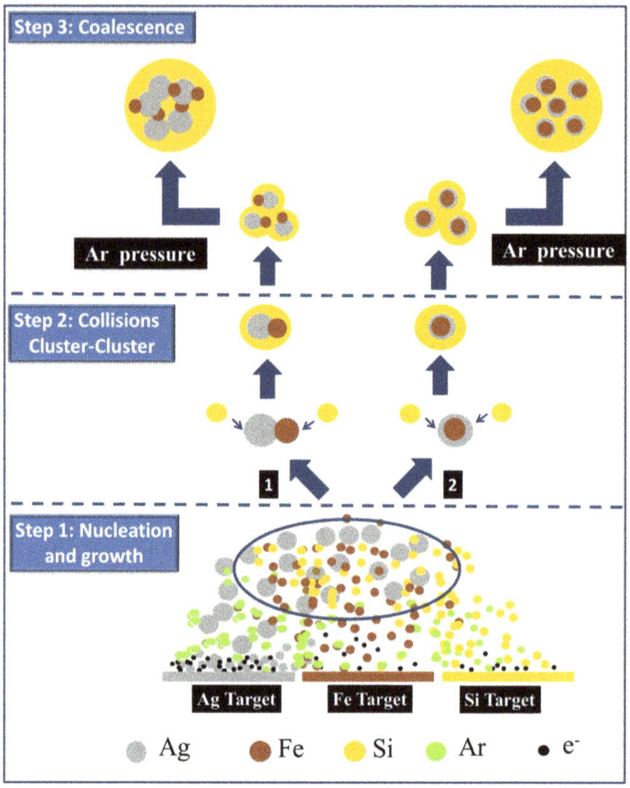

Figure 4.11. Diagram summarizing the suggested mechanism of formation of FeAg@Si MPNPs. Step 1 shows the nucleation and growth phase in the plasma zone near the surface of the sputtered targets. Step 2–1 depicts cluster–cluster collisions. Fe and Ag NPs juxtapose, forming dumbbell-like shaped NPs. Step 2–2 Ag and Fe NPs mix and then segregate in core-shell structures. Then Si nanoclusters cover the surface of FeAg NPs resulting in a core/shell structure. The resulting core/shell NPs coalesce in-flight forming multicore/shell NPs (Step 3).

In this section two examples of multielement HNPs are presented. The first example addresses the production of Au–Pt–Pd HNPs by the MS-IGC system presented in figure 2.2 [31]. The second example discusses the production of FeRh@Au intermetallic-core@metallic-shell NPs, prepared using the magnetron sputtering technique [36].

4.8.2 The synthesis of Au–Pt–Pd HNPs

In this example [31], the authors used the MS-IGC system described in figures 2.2 and 2.3. Briefly, three one-inch independent targets of Au, Pt and Pd were simultaneously co-sputtered in the aggregation chamber. The initial base pressure was $\sim 2.0 \times 10^{-8}$ mbar. During deposition process, Ar gas flow rate and aggregation zone length were set at 70 sccm and 100 mm, respectively. The pressures in the aggregation chamber and deposition chamber were set at $\sim 25 \times 10^{-2}$ and $\sim 7.0 \times 10^{-4}$ mbar,

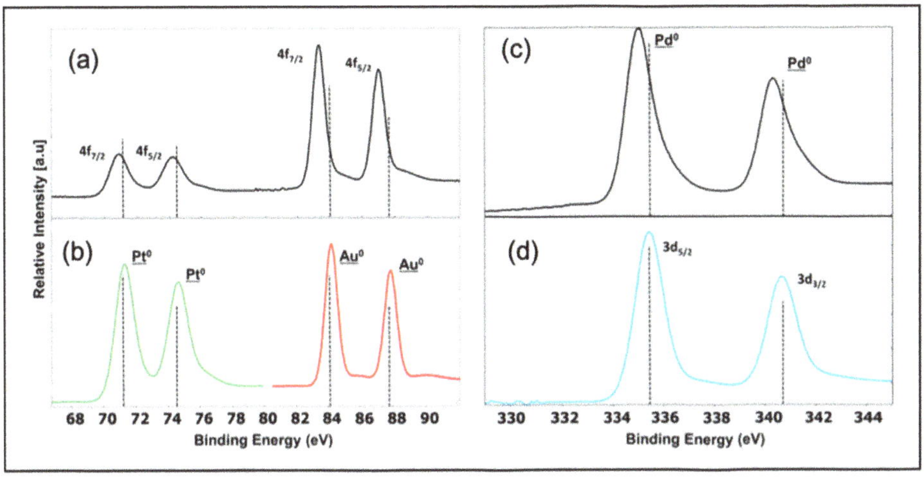

Figure 4.12. XPS spectra corresponding to (a) Pt 4f and Au 4f, (c) Pd 3d core levels from the sample Au-5W/Pt-5W/Pd-15W. (b) and (d) show reference peaks for monometallic NP samples. The dashed lines indicate the binding energies of the metal references. Reproduced with permission from [31]. Copyright ACS 2019.

respectively. Among the optimum experimental conditions to produce Au–Pt–Pd trimetallic NPs, the authors found that, the application of DC power of 5 W to Au and Pt targets, and 15 W to Pd target leads to the production of controlled and homogeneous trimetallic HNPs with a Pd shell and Pt–Au core. In fact, Pd with its faster nucleation rate forms initially monometallic nuclei, which act as seeds for Au and Pt adatoms deposition. As a result, a higher Pd sputtering power is necessary to increase the number of potential trimetallic nucleation sites. The Au-5W/Pt-5W/Pd-15W sample was analyzed using XPS, as shown in figures 4.12 [31]. In all cases the peaks corresponding to Pt 4f, Au 4f and Pd 3d core level spectra show important shift towards higher binding energies, indicating strong interaction among the three atomic species [20].

EDX analysis corresponding to a representative trimetallic HNP (sample Au-5W/Pt-5W/Pd-15W), is shown in figures 4.13. The HR-STEM image in figure 4.13(a) shows higher intensity in the core, which can be attributed to the presence of Au and/or Pt. A line scan along the dashed green line displays a compositional gradient from the core to the shell, with the Pd concentration showing a higher intensity in the shell region and a small decrease in the core region (figures 4.13(b)). Whereas the core region shows an important Au and Pt enrichment compared to the surface. The elemental mapping displayed in figures 4.13(c)–(e) confirms the Au–Pt-rich-core@Pd-rich shell configuration.

4.8.3 The synthesis of FeRh@Au HNPs

In this example [36], the authors deposited the trimetallic NPs on amorphous alumina (a-Al_2O_3) by DC magnetron sputtering from three high-purity Fe, Rh, and Au elemental targets in an ultrahigh vacuum (UHV) chamber. The a-Al_2O_3 layer

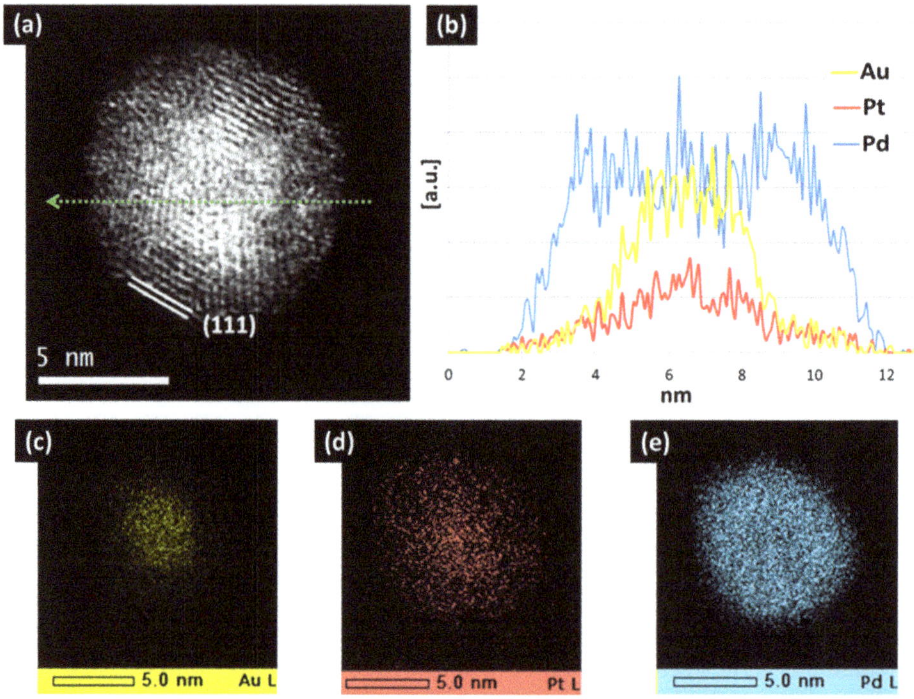

Figure 4.13. (a) HR-STEM image of a multifaceted Au-5W/Pt-5W/Pd-15W NP. (b) its corresponding EDX line scan. The green dashed line in (a) represents the path of the scanning beam. (c–e) Local elemental mapping by EDX showing the core–shell structure (the intensity corresponds to counts). Reproduced with permission from [31]. Copyright ACS 2019.

was deposited on a water-soluble NaCl crystal. To achieve a core–shell morphology, the authors designed a two-step synthesis process: (1) first, B2-FeRh NPs were grown by co-deposition from Fe and Rh targets on the a-Al_2O_3/NaCl substrate, whose temperature was brought to 750 °C; (2) the second step consisted of the deposition of a Au layer at 600 °C on the already formed B2-FeRh NPs. The sputtering conditions were tuned to near a 1:1:1 Fe/Rh/Au composition. Finally, an amorphous a-Al_2O_3 layer was deposited at room temperature to cover the NPs and thus to prevent water damage during the removal of the NaCl support. Figure 4.14 confirms the successful formation of the trimetallic NPs and the core@shell morphology through the EDX mapping of the sample.

Using a physical vapor deposition route, the authors succeeded to grow trimetallic NPs in the 10 nm range with well-controlled composition and element distribution. In particular, the FeRh core was successfully synthesized in its intermetallic B2 phase displaying a chemical order at the atomic level. The Au layer, deposited in the second step of the process, grows epitaxially on the $\{100\}_{B2}$ and $\{110\}_{B2}$ facets (figure 4.14(a)), which are the preferential surface facets of the nanocrystalline intermetallic core. Au remains at the surface of the FeRh crystals, forming a complete shell [36].

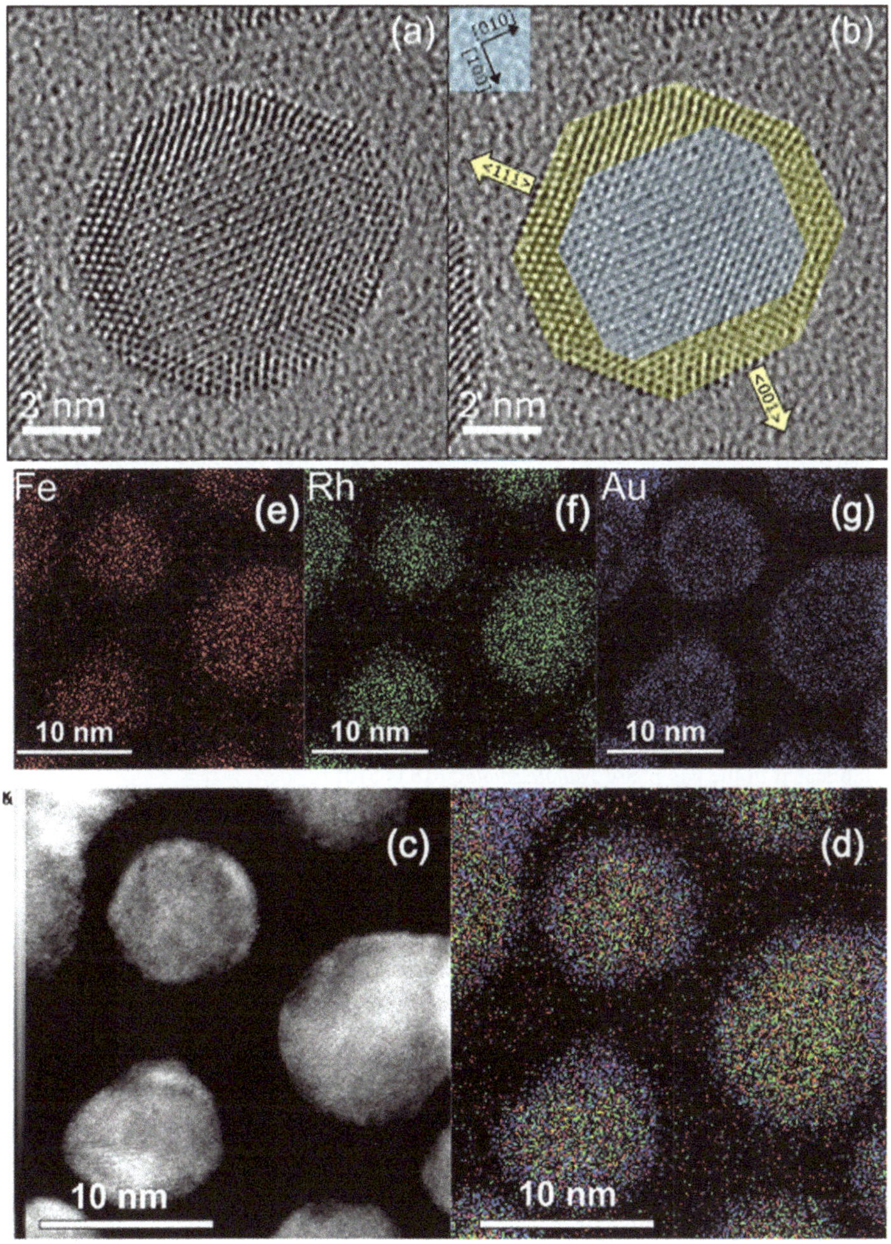

Figure 4.14. (a) HREM image of FeRh@Au NPs. (b) Artificially colored image corresponding to the image in (a). The orientations of the FeRh core and of Au shell are indicated using the same color as the corresponding regions. (c) STEM image of the FeRh@Au NPs. (d) EDX map of (c) showing the chemical composition of the NPs. (e–g) Elemental maps of Fe, Rh, and Au corresponding to the same image in (c). Adapted with permission from [36]. Copyright ACS 2020.

References

[1] Wang C, Xu C, Zeng H and Sun S 2009 Recent progress in syntheses and applications of dumbbell-like nanoparticles *Adv. Mater.* **21** 3045–52

[2] Parsina I, DiPaola C and Baletto F 2012 A novel structural motif for free CoPt nanoalloys *Nanoscale* **4** 1160–6

[3] Yu H, Chen M, Rice P M, Wang S X, White R L and Sun S 2005 Dumbbell-like bifunctional Au–Fe_3O_4 nanoparticles *Nano Lett.* **5** 379–82

[4] Gu H, Yang Z, Gao J, Chang C K and Xu B 2005 Heterodimers of nanoparticles: formation at a liquid–liquid interface and particle-specific surface modification by functional molecules *J. Am. Chem. Soc.* **127** 34–5

[5] Zhang L, Dou Y H and Gu H C 2006 Synthesis of Ag–Fe_3O_4 heterodimeric nanoparticles *J. Colloid Interface Sci.* **297** 660–4

[6] Lu Y, Xiong H, Jiang X and Xia Y 2003 Asymmetric dimers can be formed by dewetting half-shells of gold deposited on the surfaces of spherical oxide colloids *J. Am. Chem. Soc.* **125** 12724–5

[7] Meng X, Seton H C, Lu L T, Prior I A, Thanh N T K and Song B 2011 Magnetic CoPt nanoparticles as MRI contrast agent for transplanted neural stem cells detection *Nanoscale* **3** 977–84

[8] Zeng H and Sun S 2008 Syntheses, properties, and potential applications of multicomponent magnetic nanoparticles *Adv. Funct. Mater.* **18** 391–400

[9] Perro A, Reculusa S, Ravaine S, Bourgeat-Lami E and Duguet E 2005 Design and synthesis of Janus nanoparticles *J. Mater. Chem.* **15** 3745–60

[10] Bai J, Xu Y H, Thomas J and Wang J P 2007 $(FeCo)_3Si$–SiO_x core–shell nanoparticles fabricated in the gas phase *Nanotechnology* **18** 065701

[11] Xu Y H and Wang J P 2008 Direct gas-phase synthesis of heterostructured nanoparticles through phase separation and surface segregation *Adv. Mater.* **20** 994–9

[12] Skumryev V, Stoyanov S, Zhang Y, Hadjipanayis G, Givord D and Nogués J 2003 Beating the superparamagnetic limit with exchange bias *Nature* **423** 850–3

[13] Kaur M, McCloy J S, Jiang W, Yao Q and Qiang Y 2012 Size dependence of inter- and intracluster interactions in core–shell iron–iron oxide nanoclusters *J. Phys. Chem.* C **116** 12875–85

[14] Sumiyama K, Katoh R, Kadowaki S and Hihara T 2010 Fe–Si core/Si-shell clusters prepared by double glow discharge sources *J. Nanopart. Res.* **12** 2589–96

[15] Tanaka N, Sumiyama K, Katoh R, Hihara T, Sato K, Konno T J and Mibu K 2010 Core-shell formation and juxtaposition in Fe and Si hybrid clusters prepared by controlling the collision stages *Mater. Trans.* **51** 1990–6

[16] Elsukova A, Li Z A, Möller C, Spasova M, Acet M, Farle M, Kawasaki M, Ercius P and Duden T 2011 Structure, morphology, and aging of Ag–Fe dumbbell nanoparticles *Phys. Status Solidi* A **208** 2437–42

[17] Benelmekki M, Bohra M, Kim J H, Diaz R E, Vernieres J, Grammatikopoulos P and Sowwan M 2014 Facile single-step synthesis of ternary multicore magneto-plasmonic nanoparticles *Nanoscale* **6** 3532–5

[18] Benelmekki M, Vernieres J, Kim J H, Diaz R E, Grammatikopoulos P and Sowwan M 2015 On the formation of ternary metallic-dielectric multicore-shell nanoparticles by inert-gas condensation method *Mater. Chem. Phys.* **151** 275–81

[19] Yamashita T and Hayes P 2008 Analysis of XPS spectra of Fe^{2+} and Fe^{3+} ions in oxide materials *Appl. Surf. Sci.* **254** 2441–9
[20] Moulder J F, Stickle W F, Sobol P E and Bomben K D 1992 *Handbook of X-ray Photoelectron Spectroscopy* ed J Chastain (Waltham, MA: Perkin Elmer)
[21] Nanbu K and Kondo S 1997 Analysis of three-dimensional DC magnetron discharge by the particle-in-cell/Monte Carlo method *Jpn. J. Appl. Phys.* **36** 4808–14
[22] Ido S, Kashiwawagi M and Takahashi M 1999 Computational studies of plasma generation and control in a magnetron sputtering system *Jpn. J. Appl. Phys.* **38** 4450–4
[23] www.iap.tuwien.ac.at/www/surface/sputteryield, A simple sputter yield calculator, accessed June 2014
[24] Hara S, Izumi S, kumagai T and Sakai S 2005 Surface energy, stress and structure of well-relaxed amorphous silicon: a combination approach of *ab initio* and classical molecular dynamics *Surf. Sci.* **585** 17–24
[25] Benelmekki M and Sowwan M 2015 *Handbook of Nano-Ceramic and Nano-Composite Coatings and Materials* ed H Makhlouf and D Scharnweber (Amsterdam: Elsevier) ch 9 207–23
[26] Yu W, Porosoff M D and Chen J G 2012 Review of Pt-based bimetallic catalysis: from model surfaces to supported catalysts *Chem. Rev.* **112** 5780–817
[27] Peng Z and Yang H 2009 Designer platinum nanoparticles: control of shape, composition in alloy, nanostructure and electrocatalytic property *Nano Today* **4** 143–64
[28] Zhang L, Iyyamperumal R, Yancey D F, Crooks R M and Henkelman G 2013 Design of Pt-shell nanoparticles with alloy cores for the oxygen reduction reaction *ACS Nano* **7** 9168–72
[29] Li L, Yifeng E, Yuan J, Luo X, Yang Y and Fan L 2011 Electrosynthesis of Pd/Au hollow cone-like microstructures for electrocatalytic formic acid oxidation *Electrochim. Acta* **56** 6237–44
[30] Ahmed M, Attard G A, Wright E and Sharman J 2013 Methanol and formic acid electrooxidation on Nafion modified Pd/Pt{111}: the role of anion specific adsorption in electrocatalytic activity *Catal. Today* **202** 128–34
[31] Mattei J-G *et al* 2019 Gas-phase synthesis of trimetallic nanoparticles *Chem. Mater.* **31**(6) 2151–63
[32] Lacey S D *et al* 2019 Stable multimetallic nanoparticles for oxygen electrocatalysis *Nano Lett.* **19** 5149–58
[33] Chen P-C *et al* 2015 Tip-directed synthesis of multimetallic nanoparticles *J. Am. Chem. Soc.* **137** 9167–73
[34] Li J and Sun S 2019 Intermetallic nanoparticles: synthetic control and their enhanced electrocatalysis *Acc. Chem. Res.* **52** 2015–25
[35] Wang D, Xin H and Hovden R *et al* 2013 Structurally ordered intermetallic platinum–cobalt core–shell nanoparticles with enhanced activity and stability as oxygen reduction electrocatalysts *Nature Mater* **12** 81–7
[36] Benzo P, Combettes S, Garcia C, Hungria T, Pecassou B and Casanove M-J 2020 Epitaxial growth of a gold shell on intermetallic FeRh nanocrystals *Cryst. Growth Des.* **20** 4144–9

IOP Publishing

Designing Hybrid Nanoparticles (Second Edition)

Maria Benelmekki

Chapter 5

Obtaining anisotropic magnetic nanostructures

The assembly of nanoparticles (NPs) to produce 1D and 2D nanostructures is essential to build tailored nanostructures for applications ranging from nanoelectronics to nanomedicine. For example, assembling magnetoplasmonic nanostructures enables multimodal biosensing and bioimaging for early diagnosis and prognosis of diseases. However, to date, the synthesis of such complex nanostructures is mostly relying on solution-based methods. These methods usually require complex, time-consuming, and often costly synthesis routes. This chapter introduces two examples of how gas-phase methods can be employed for the successful magnetic-assisted assembly of Fe nanoclusters (NCs) and NPs, to form stable and permanent 1D and branched Fe nanostructures. The first example addresses the formation of in-flight 1D Fe and FeAg nanorods (NRs) using a conventional modified MS-IGC, before reaching the deposition substrate. The second example addresses the formation of Fe and Ni/NiO nanowires and nanotrusses while applying a magnetic field on the surface of the deposition substrate.

5.1 Introduction to the self-assembly of magnetic NPs

Assembling inorganic nanomaterials with different modalities has created new paradigms for a wide range of applications within the fields of energy, healthcare, and chemical-production [1–5]. Various anisotropic assemblies, such as one-dimensional (1D) wires and rods, two-dimensional (2D) disks and prisms, and branched shapes have been investigated as potential nanobuilding blocks for nanostructures and nanodevices [6–10]. However, to date, solution-based syntheses are the most common routes to produce such advanced nanostructures, which usually require complex, time-consuming, and often costly solution-based synthesis routes. Specifically, the synthesis routes of magnetic and magnetoplasmonic materials with controllable compositions, sizes, and shapes are still relying on additives, such as stabilizer ligands, organic surfactants, and templates [11–17]. Gas-phase methods are attractive alternatives to produce shape-controlled magnetic nanostructures. They enable the synthesis of

doi:10.1088/978-0-7503-3587-4ch5

various metal nano-dimensional objects. In fact, the possibility to produce these nanostructures without the use of solvents and/or templates is an important step towards an environmentally friendly production of nanostructures free of any residual impurities [8, 18]. Further information and examples are discussed in chapter 2. In this chapter, we present two examples of the production of 1D and branched magnetic nanostructures produced by two different gas-phase systems. The first example addresses the formation of in-flight Fe and FeAg NRs using a conventional modified MS-IGC, as shown in figure 2.5. In this case, the formed NRs were subjected to a magnetic field during their pathway towards the deposition chamber. The second example addresses the formation of Fe nanowires and nanotrusses using the gas-phase system presented in figure 2.9. In this case a hollow Fe cathode was used as a source of Fe atoms, and a the magnetic field was applied directly to the deposition substrate.

5.2 The synthesis and characterization of Fe and FeAg NRs

5.2.1 Deposition of Fe and FeAg NRs

NRs were synthesized using the MS-IGC method as described in figure 2.5 [19]. For the production of Fe NRs, three optimized Fe targets of one inch of diameter were simultaneously DC sputtered (figure 2.5(a)). The power applied to each target were 10 W (~2 W cm^{-2}), and the Ar flux in the aggregation-chamber was set at 80 sccm. The experimental parameters are summarized in table 5.1.

The generated vapor condenses into NCs along its propagation path through the aggregation-chamber, whose length was set at 90 mm. These NCs are extracted and accelerated by the differential pressure between the aggregation-chamber and the shell-coater chamber. In the shell-coater chamber, a flux of Ar of 10 sccm was introduced resulting in an estimated working pressure of ~4.3×10^{-2} mbar, and no power was applied to the Ag-target. The nanostructures were extracted to a deposition chamber, where they were deposited on flat silicon substrates (5 × 5 mm^2) and carbon-coated copper TEM grids.

For the synthesis of FeAg NRs, the above described experimental conditions were maintained, and a custom-shaped Ag target (150 mm length × 50.8 mm width × 3.0 mm thick), placed in the shell-coater chamber, was DC sputtered with an Ar gas flux of 10 sccm at 50 W of power (table 5.1). The nanostructures are extracted from one chamber to another by the differential pressure between the three chambers.

Table 5.1. Experimental conditions for NR deposition.

	Aggregation chamber (reading pressure)	Shell-coater chamber (estimated pressure)	Deposition chamber (reading pressure)
Base pressure (mbar)	1.5×10^{-6}	1.5×10^{-6}	8.7×10^{-8}
Working pressure (mbar)	3.5×10^{-1}	4.3×10^{-2}	1.3×10^{-5}
Power applied to the targets (W/cm^2)	~2 (Fe target)	~0.65 (Ag target)	—

The resulting dragging forces are estimated based on the chamber pressures and the size of the formed NPs (see table 5.2).

5.2.2 Characterization methods

AFM (Multimode 8, Bruker, CA) was used for morphological characterization of the nanostructures. SEM images were obtained using a FEI-Helios NanoLab 650. STEM/EDX analyses were performed using a FEI Titan 80–300 kV operated at 300 kV. The in-plane field dependence of magnetization (M (H)) was measured using a Quantum Design cryogen-free physical property measurement system (PPMS).

5.2.3 Morphology and composition of Fe and FeAg NRs

A low magnification SEM image (figure 5.1) shows the anisotropic shape of the obtained Fe NRs. These nanostructures show random orientations and irregular aspect ratios. Figure 5.2 presents tapping mode AFM images of the Fe NRs, confirming the anisotropic shape of the obtained nanostructures. The height of the different nanostructures is homogeneous, of about 6 nm. Different shapes are observed encompassing elongated NRs, 'Y' and 'V' shaped nanostructures

Table 5.2. Dragging forces in the aggregation-chamber and shell-coater chamber.

NP diameter (nm)	2.5	5	10
$F_{\mathrm{drag0}} \times 10^{-18}$ (N)	~602.50	~2410	~9649
$F_{\mathrm{drag}} \times 10^{-18}$ (N)	~84.37	~337.50	~1350

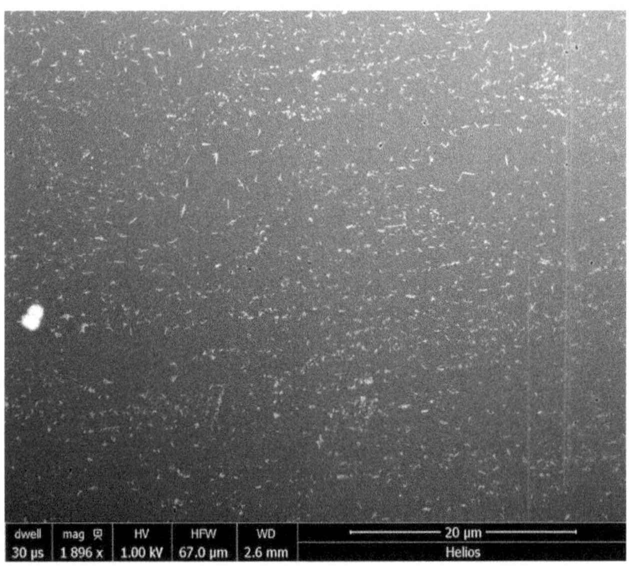

Figure 5.1. Panoramic low-magnification SEM image showing the as-deposited Fe NRs on Si substrate.

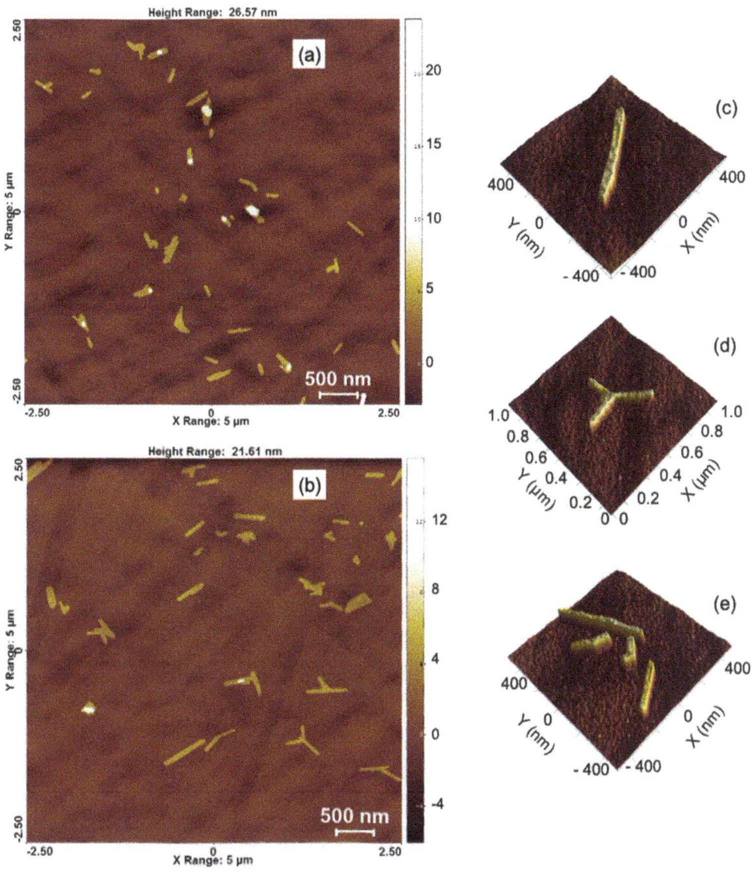

Figure 5.2. (a) and (b) A panoramic tapping mode AFM images of the Fe-NRs. (c)–(e) 3D images of representative branched Fe NRs.

(figure 5.2(c)–(e)). The V-shaped nanostructures presumably result from two NRs connected via end-to-end mode to eliminate two high-energy surfaces [15]. The Y-shaped nanostructures comprise three segments like one stem with two branches. A plausible explanation of the formation of Y-shaped NRs with different widths of stem and branches is the combination of a V-shaped NR with a straight one via a side-to-side contact mode, to enhance the packing density [15, 19, 20]. Hence, the formation of V and Y morphologies of Fe NRs is likely to be driven by the need of the formed nanostructures to lower their overall energy.

To determine the chemical composition of the NRs, EDX analyses were conducted. Figure 5.3(a) shows the EDX spectrum corresponding to the position denoted as (1) in the STEM image in (b), which shows the image of an individual NR deposited on a conventional TEM Cu-grid. The yellow rectangle indicates the region where the electron beam was focused for the acquisition of the data. The spectrum confirms the chemical composition of the Fe NR. No Ag chemical component is observed (Cu chemical component originates from the TEM grid).

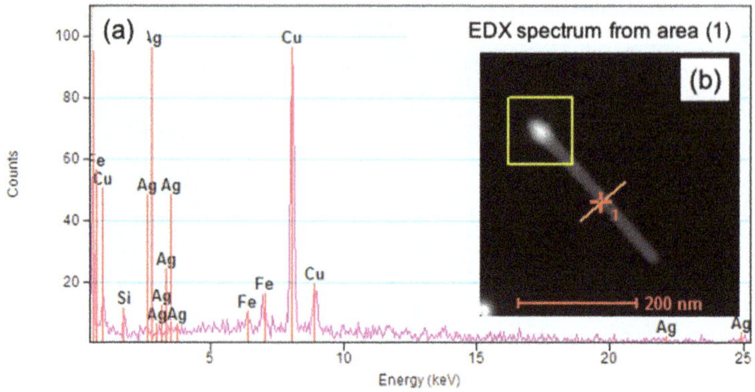

Figure 5.3. (a) EDX spectrum corresponding to position (1) indicated in (b). (b) STEM image of a Fe NR deposited on a Cu-grid.

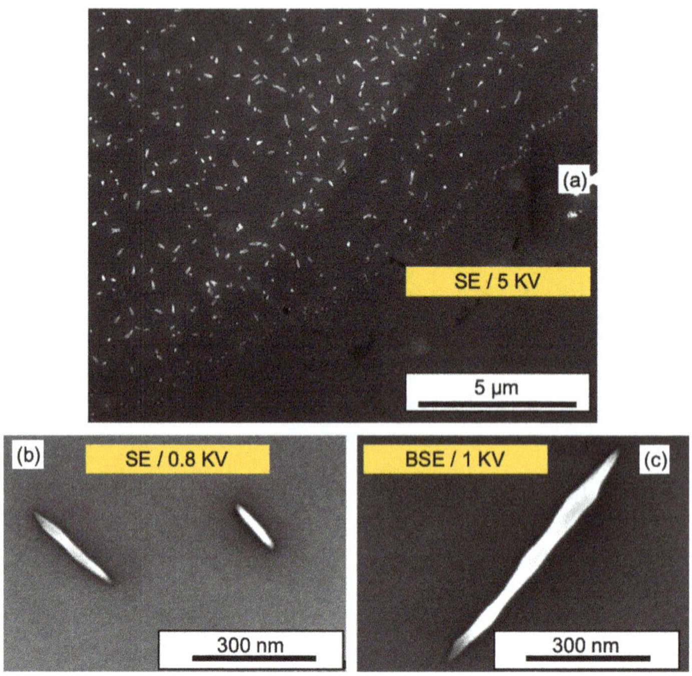

Figure 5.4. (a) Low-magnification SEM image of FeAg NRs. (b) and (c) higher-magnification SEM images of FeAg NRs collected in SE and BSE modes, respectively.

Figures 5.4(a)–(c) show the SEM images of the obtained nanostructures, resulting from the simultaneous sputtering of Fe-targets in the aggregation-chamber, and the Ag-target in the shell-coater chamber. Figure 5.4(a) reveals the formation of elongated nanostructures with different aspect ratios. The observed waved-like surface in (b) and (c) is likely to occur as a consequence of the neck formation during the coalescence process of Fe-NCs, which will be discussed in section 5.2.4.

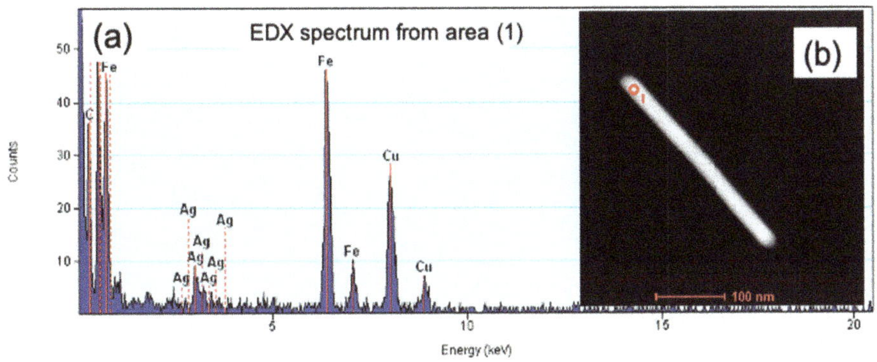

Figure 5.5. (a) EDX spectrum from position (1) in (b) showing the chemical composition of the NR in (b). (b) STEM image of an individual FeAg NR.

No evidence of bundle formation or branching are observed. To determine the chemical composition of these NRs, EDX analysis was conducted (figures 5.5(a) and (b)). The resulting spectrum shows the presence of both Fe and Ag components (Cu originates from TEM grid).

5.2.4 Effects of the vacuum conditions during SEM imaging

Regarding the challenges commonly faced during surface analysis of nanomaterials, this section brings to the light the importance of the vacuum conditions during SEM characterization, through the contrast-reversal observed in the SEM images corresponding to the Fe NRs.

Figure 5.6(a) shows a low-magnification SEM image of the Fe NRs. A higher magnification SEM image (~7.5 times) of the denoted zone (1) in figure 5.6(a) reveals the formation of different shapes (figure 5.6(b)): (i) the alignment of the NRs in a side-to-side mode to form bundles randomly distributed on the surface of the substrate, as indicated with the white arrows; (ii) a V-shaped NR with inhomogeneous diameter is shown in figure 5.6(c), corresponding to a higher magnification SEM image (~23 times) of the denoted zone (2) in figure 5.6(b). As discussed in the previous section, these rearrangements are likely to be driven by the need to reduce the overall surface energy [7, 21].

In comparison to SEM images of FeAg NRs (figure 5.4), the waved-like surface is more noticeable for FeAg NRs due to the higher electrical conductivity of Ag component, enabling a better resolution of the SEM images.

Remarkably, the SEM images show a shift in the color contrast of the NRs from bright contrast (figure 5.6(a)) to dark contrast (figures 5.6(b) and (c)), when performing higher magnifications. This contrast-reversal is observed at different accelerating voltages, in both secondary electrons (SE) and backscattered electrons (BSE) detection modes. However, regardless of the accelerating voltage or the detection mode, such contrast-reversal is not observed in FeAg NRs (figure 5.4). Of note, during the SEM imaging, all the other experimental parameters were maintained unchanged, so that contrast-reversals originating from detection angle

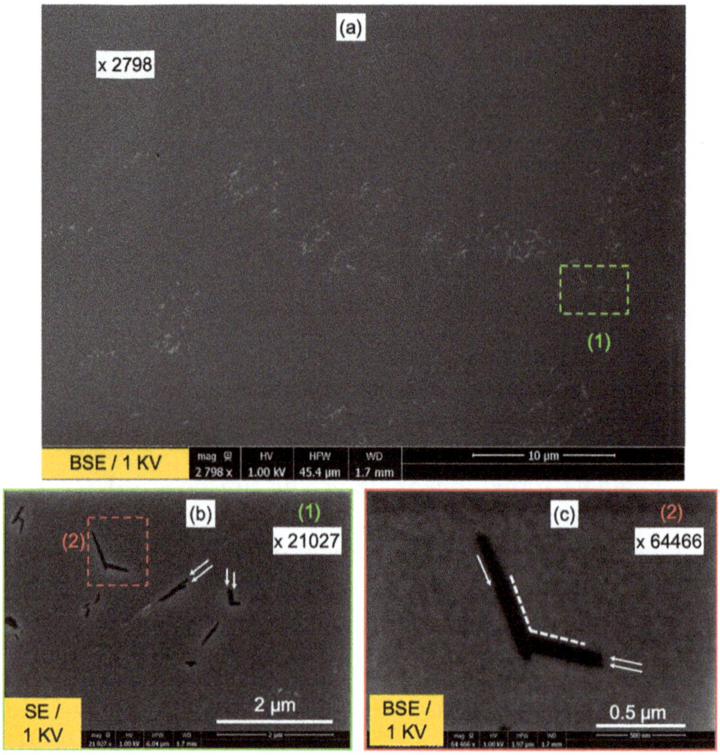

Figure 5.6. (a) Low-magnification SEM image of the Fe-NRs. (b) and (c) higher-magnification SEM images corresponding to region (1) in (a) and region (2) in (b), respectively.

or sample position/orientation can be discarded [22]. Hence, this alteration of the color–contrast is likely to be related to the vacuum conditions within the measuring-chamber of the SEM (~5 $\times 10^{-6}$ mbar). In fact, high-vacuum (~10^{-8} mbar) and ultrahigh-vacuum (<10^{-9} mbar) are crucial to conduct sensitive surface analysis of materials, and vacuum conditions are even more critical for nanomaterials, due to their high reactivity with the surrounding environment. Hence, for Fe NRs, when the SE and BSE are ejected from the sample, they are likely to interact with the residual oxygen near the surface of the Fe NRs. Thus, the resulting dissociated O-atoms react with Fe-atoms to form FeO_x component, which reduces the electrical conductivity of the NRs [19].

In fact, XPS studies performed on Fe NRs and FeAg NRs (after their exposure of the samples to the atmosphere) showed the presence of metallic Fe component in FeAg NRs, which suggests that the Ag component forms either a thin film or inhomogeneous coating on the surface of the Fe NRs providing their partial protection against oxidation [4, 19, 23], in agreement with EDX results shown in figure 5.5. Further discussion on the position of each component in FeAg NRs is given in the following section.

In summary, due to the high-reactivity of nanomaterials to their surrounding medium, the formation of agglomerates is the most common mechanism for free-standing NPs to reduce their surface energy. When the sample is exposed to the atmosphere, the formed agglomerates (and even the individual NPs) are likely to be released from the surface of the supporting substrate to the surrounding environment, resulting in an important decrease of the remaining amount of the sample to be analyzed, which might affect the quality of the acquired data. Therefore, special attention needs to be paid during the manipulation, preparation and transfer of free-standing samples, so that *size effects* can be minimized, and thus accurate and high quality data can be achieved, specially for surface sensitive techniques such as EM and XPS [9, 24, 25].

5.3 The Mechanism of the formation of Fe and FeAg NRs

When the Fe targets are DC-sputtered (figure 2.5(a)), nucleation and growth of tiny Fe NCs occur during their flight within the aggregation chamber. Since the surfaces of these NCs are clean and not oxidized, and the fraction of surface atoms per unit volume is very large (size effects), the Fe NCs continue to collide and coalesce with each other by diffusion of the constituent atoms at their contact interfaces, resulting in the formation of larger amorphous NPs [1, 19, 23].

The NPs entering the shell-coater chamber (figure 2.5(b)) will experience electrostatic repulsion, Van der Waals interaction, dragging force (F_{drag}) driven by the pressure differential between the core–shell chamber and the deposition chamber, and the magnetic force (F_{mag}) resulting from the magnetic field gradient, which originates from the magnetron sputtering cathode (figure 5.7). Following the DLVO

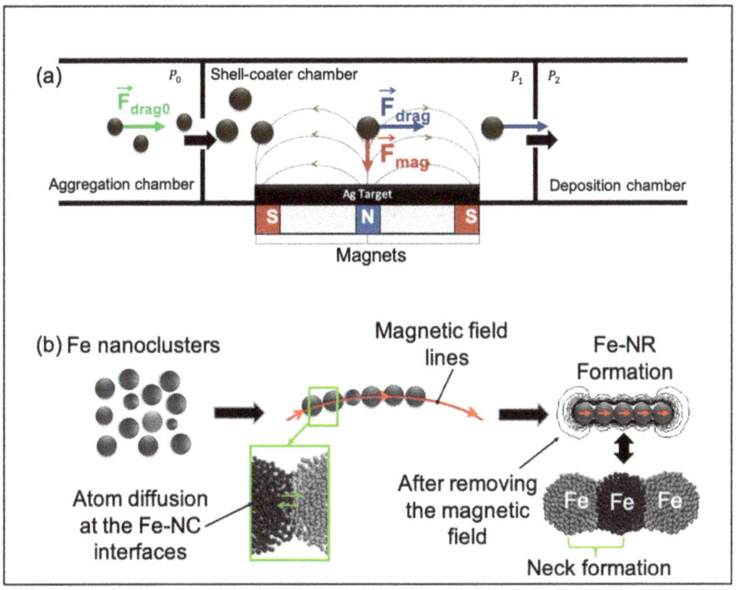

Figure 5.7. (a) A scheme illustrating the moving of the NPs through the different chambers. F_{mag} and F_{drag} indicate the magnetic and dragging forces acting on the NPs, respectively. (b) Schematic representation of NR formation.

theory, both electrostatic repulsion and Wan der Waals attraction, depend on the particle-to-particle distance, and play an important role at very short distances [25]. Considering the low coverage of NPs obtained from this MS-IGC system, it is assumable that the concentration of the NPs is low enough, and thus both Van der Waals attraction and electrostatic repulsion become small, and even negligible [19]. Therefore, the dragging force and magnetic force are likely to be the forces playing a significant role on the behavior of the Fe NPs within the shell-coater chamber (figure 5.7).

5.3.1 Estimation of the dragging force acting on the Fe NPs

Based on the pressure difference between the shell-coater chamber and the deposition chamber, an isolated spherical Fe NP (or NC) of a diameter D, will experience dragging forces within the aggregation chamber (F_{drag0}) and the shell-coater chamber (F_{drag}), which can be approximated as follows [26]:

$$F_{\text{drag0}} = S \cdot \Delta P_{\text{Ag}}, \tag{5.1a}$$

$$F_{\text{drag}} = S \cdot \Delta P_{\text{SC}}, \tag{5.1b}$$

where $S = \pi D^2$ (in m^2) is the surface of a spherical NP; $\Delta P_{\text{Ag}} = P_1 - P_0$ is the pressure differential between the aggregation chamber (P_0) and shell-coater chamber (P_1); and $\Delta P_{\text{SC}} = P_2 - P_1$ is the pressure differential between the shell-coater chamber (P_1) and deposition chamber (P_2). All the pressures are expressed in N m^{-2} (figure 5.7(a)). Table 5.2 presents the estimated values of the dragging force acting on NPs with different diameters.

5.3.2 Estimation of the magnetic force within the shell-coater chamber

On the other hand, an isolated Fe NP within the shell-coater chamber with a diameter D, will experience a magnetic force (F_{mag}) in addition the dragging force (F_{drag}), as indicated in figure 5.7(a). This magnetic force results from the magnetic field gradient originating from the set of magnets inside the cathode of the linear magnetron [19]. The calculated magnetic field gradient on the surface of the cathode is shown in figure 2.5(e).

Generally speaking, the magnetic force acting on a small magnetic particle (NP in our case) can be described by the following equation [27]:

$$F_{\text{mag}} = (m \cdot \nabla)B \tag{5.2}$$

where m is the magnetic moment (A m^2) and ∇B is the magnetic field gradient (T m^{-1}). The magnetic moment (m) is related to the particle size as follows [28]:

$$m = MV \tag{5.3}$$

where M is the magnetization of the particle (A m^{-1}), and V is the volume of a spherical particle (m^3). Therefore, equation (5.2) becomes:

$$F_{\text{mag}} = MV\nabla B = M\frac{4}{3}\pi(D/2)^3\nabla B = M\frac{\pi}{6}D^3\nabla B \quad (5.4)$$

Using equation (5.4), and the calculated magnetic field gradient values (figure 2.5(e)), table 5.3 presents an estimation of F_{mag} at different distances from the surface of the linear cathode (Ag-target in figure 5.7(a)). The magnetization M of the Fe NPs was approximated to the saturation magnetization value of pure iron ($\sim 1.75 \times 10^6$ A m^{-1}) [29].

In all cases, the magnetic force shows higher values near the surface of the target, within the first ~10 mm, and decreases to graze zero for distances > to ~40 mm, depending on the size of the NCs. The magnetic force experienced by an NP is proportional to $\sim D^3$, (increases with the size of the NPs), and thus the magnetic force experienced with a bigger NP is stronger when compared to the magnetic force experienced by a small NP. Therefore, the bigger NPs spatially close to the surface of the cathode (target), will likely undergo a larger residence time within the shell-coater chamber, which increases the probability of these NPs to collide with each other. While the motion of the small NPs is likely to be dominated by the dragging force.

On the other hand, the estimated dragging force acting on isolated Fe NPs within the shell-coater chamber is, at least, two magnitudes higher than the magnetic force, which indicates that the Fe NPs will not be trapped on the surface of the target, and thus will be transported to the deposition chamber.

5.3.3 Magnetic interaction between Fe NPs

Referring to the NP–NP magnetic interactions within the shell-coater chamber, the Fe NPs interact and rearrange to form small chains, resulting in the formation of NRs, as demonstated in SEM, STEM and AFM analyses (figure 5.7(b)). The quantification of Fe NPs' magnetic strength within the core–shell chamber, depends

Table 5.3. Calculated F_{mag} at different distances from the surface of the linear cathode (Ag-target) for NPs with different diameters (D).

		$\sim F_{\text{mag}} \times 10^{-20}$ (N)		
Distance (mm)	Grad B (T m^{-1})	$D = 2.5$ nm	$D = 5$ nm	$D = 10$ nm
5	5.90	0.09	0.94	5.43
10	5.18	0.08	0.85	4.76
15	3.50	0.06	0.62	3.53
20	2.05	0.03	0.40	1.97
25	1.20	0.02	0.21	1.12
30	0.83	0.01	0.14	0.77
35	0.62	0.009	0.11	0.57
40	0.32	0.006	0.06	0.35
45	0.062	0.001	0.01	0.08

on the volume fraction of the NPs dispersed in the shell-coater chamber on one hand; and on the ratio between the magnetic energy associated to the dipole–dipole attraction and the thermal energy, on the other hand [5, 30]. In the MS-IGC system used for this example, the experimental quantification of both parameters is not straightforward, as it requires precise information on the size distribution and concentration of the Fe NPs within the shell-coater chamber, and a better control of the magnetic force [31].

However, if we analyze the magnetization curve *M(H)* in figure 5.8(b), the NRs reach the saturation magnetization (M_s) around 1500 Oe (0.15 T), in addition *M(H)* shows a fast increment in the range 0 to ~500 Oe (0.05 T), reaching almost 50% and 25% of M_s at ~400 Oe (0.04 T) and ~100 Oe (0.01 T), respectively. If we compare these values with the calculated magnetic field in figure 5.8(c), the NPs within the first 20 mm are likely to show larger dipole–dipole interactions, and thus will interact with each other. Bigger particles might interact with each other even further from the surface of the target, as the magnetic force is proportional to D^3.

5.3.4 Interactions between Fe NPs in close contact

When the Fe NPs enter the shell-coater chamber, the magnetic field lines arising from the linear magnetron cathode (figure 5.7) induce a magnetic moment in the Fe NPs.

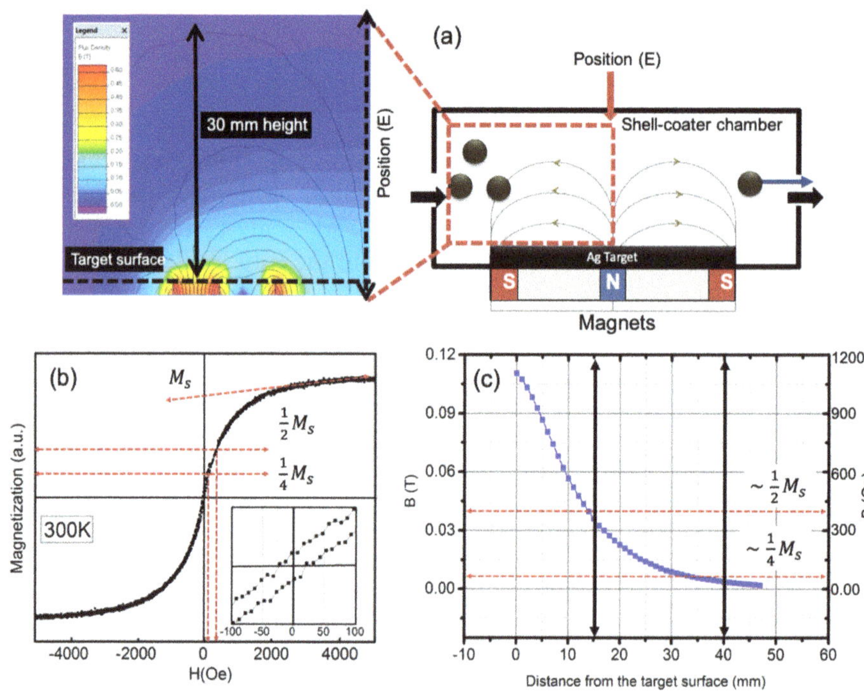

Figure 5.8. (a) Schematic showing the distribution of the magnetic field on the surface of the target. The magnetic field map is obtained from the vertical cross section (along the short side) of the linear cathode. (b) Magnetization curve *M(H)* of the Fe NRs at 300 K. (c) Profile of the calculated magnetic field from position (E) in (a). Adapted from [19], copyright (2021) with permission of Elsevier.

As discussed previously, depending on the size of the NPs and their position with respect to the surface of the linear cathode, they will tend to align their magnetic moments in the direction of the local magnetic field, giving rise to a specific directionality of interaction, and enabling Fe NPs to form small chains along the magnetic field lines (figure 5.7(b)) [7, 31, 32]. Within the shell-coater chamber, the main interactions between close Fe NPs are dipole–dipole magnetic interactions, on one hand, and exchange interactions through the surface of the NPs in close contact, on the other hand. The dipole–dipole magnetic interactions act over longer distances, inducing the agglomeration of the NPs. Thus, when two Fe NPs become close enough to make a direct intimate contact, surface diffusion driven by the need to minimize their overall free energy, takes place [21, 33]. Heat is released due to free surface annihilation, and the interface melts momentarily creating a bridge (or neck), which becomes thicker as more atoms near the interface can interact with each other (figure 5.7(b)) [34, 35]. The coalescence process continues via surface atom diffusion until the NP boundaries have vanished, and elongated shapes are created [19].

5.3.5 The formation of FeAg NRs

Ag atoms are sputtered from the linear magnetron sputtering target, positioned parallel to the flight-direction of the Fe NRs. Hence, Fe NRs act as seeds for Ag condensation and nucleation. The interaction of Ag atoms with the surface atoms of the Fe NRs, would result in the formation of an interface Fe/Ag upon solidification [8]. In fact, according to Cassidy *et al*, where the authors reported on the synthesis of Si NPs and their Ag-coating, using the same MS-IGC system [36], Ag vapor will not coalesce into nanoclusters in free space within the shell-coater chamber, and will condense on the surface of the Si NPs, as shown in figure 5.9. For more information of the nucleation and condensation process of Ag atoms, see [36]. Therefore, in the case of the FeAg NRs, the Ag atoms are expected to condense, nucleate and diffuse into the Fe NRs, forming an intermediate alloying stage, and then segregate upon solidification, which is consistent with the immiscibility of Ag and Fe in the solid

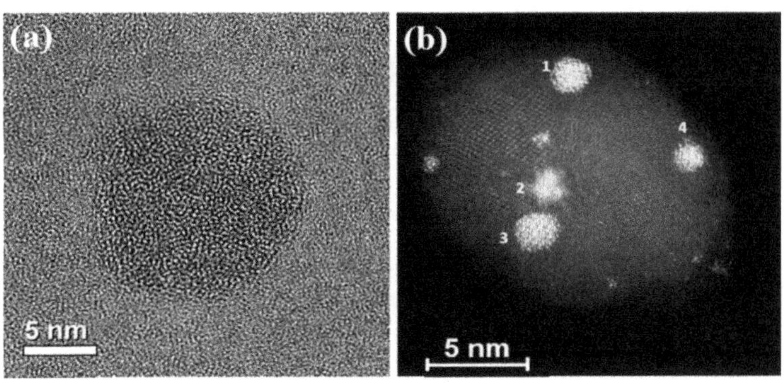

Figure 5.9. (a) TEM image of amorphous Si NPs produced by MS-IGC without Ag coating. (b) HAADF-STEM image of coated Si NP decorated with several Ag nanoclusters (bright) marked by numbers 1 to 4, respectively. Reprinted with permission from Springer Nature [36]. Copyright 2013.

state. The resulting Ag islands on the surface of Fe would have a size of about 3 nm and will continue to grow in size and number as the process continues, which might result either in a very thin and uniform film of Ag, or in a discontinuous thin film formed of Ag islands [8, 36].

5.4 The production of magnetic nanowires and nanotrusses

The emphasis of this section is to introduce a novel approach for the production of magnetic NPs, using a combination of a pulsed-plasma process with the subsequent application of an external magnetic field [37, 38]. The authors demonstrated that this novel approach results in an efficient production of supported ordered and stable ferromagnetic nanostructures on a substrate of choice.

5.4.1 Fe nanowires and nanotrusses

For this purpose, the authors used a hollow Fe cathode sputtering in a high-vacuum system with a base pressure of ~3.9×10^{-7} mbar. More details on the instrument were introduced in chapter 2. Briefly, the cathode, with an inner diameter of 5 mm and a length of 54 mm, was mounted on a copper block and placed on a magnetron to provide electrical connection and water cooling. The magnetron magnets were previously removed. A cylindrical mesh was placed in front of the cathode to confine the plasma. A grounded anode in the form of a ring, was mounted as indicated in figure 5.10(a). the argon gas was introduced through the hollow cathode with a flow rate of 70 sccm. the working pressure was set at ~1.06 mbar during the deposition. The electric power applied to the cathode was supplied using a pulsed power supply, providing a pulsing frequency of 1.2 kHz and the pulse width at 80 μs. The average output power was kept constant, at 75 W. The blue shaded area in figure 5.10(a) represents the high electron temperature region, and the dotted red line indicates the outer edge of the magnetic capture zone for 40 nm NPs [38]. To collect the sample, a permanent magnet was attached to the holder of the collecting substrate (see also figure 2.9), and the magnetic flux density at the center of the substrate surface facing the plasma was found to be 0.36 T. The experimental details can be seen in [38].

The authors showed that branched wires of Fe NPs can be fabricated onto both metal and carbon-paper substrates and that the nanostructure stability is due to both strong particle–particle bonding and a large degree of interlinking between the nanowires through branching [38]

5.4.2 Ni/NiO nanowires and nanotrusses

The authors studied other materials such as Ni/NiO nanostructures, using the same method [37]. In this case, a hollow Ni cathode (99.98% purity; inner diameter 5 mm and length 54 mm) was used to produce the nanowires and nanotrusses. The authors discussed the effects of the presence of NiO in Ni/NiO NPs in the formation of the nanowires and nanotrusses. A strong correlation between the amount of magnetic Ni and the final size and shape of the nanotrusses was observed. The decrease of the Ni amount results in the decrease of the length of the individual nanowires within the trusses, together with a higher degree of branching of the nanowires. For lower

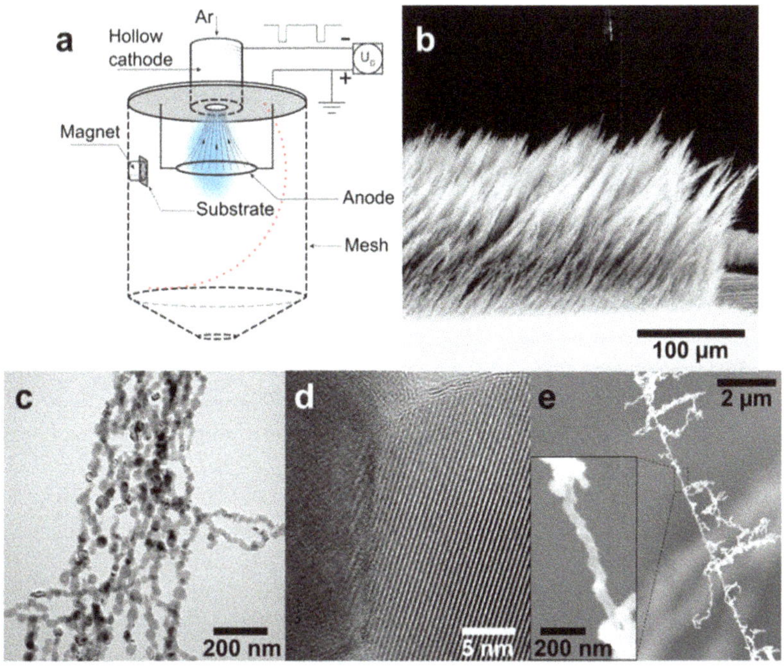

Figure 5.10. (a) A diagram of the pulsed-plasma setup. (b) SEM image of a dense array oriented by an external magnet. (c) TEM image showing the architecture of individual strains in the array in (b). (d) TEM image of the contact between two NPs. (e) SEM image of a stretched nanowire. Reprinted with permission from [38]. Copyright (2018) American Chemical Society.

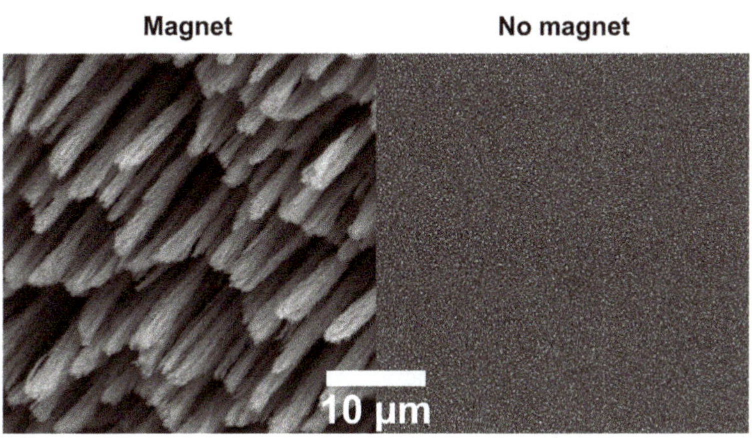

Figure 5.11. SEM images showing the effect of attaching a magnet to the substrate where the Ni/NiO NPs are deposited. Adapted from [37] with permission of Springer under CC BY 4.0.

concentrations of NiO phase, the electrically conducting Ni nanowire framework is maintained through the preferential bonding between the Ni crystals. For larger amounts of NiO phase, the Ni nanowire framework is interrupted by the NiO. These findings open the possibility for using not only magnetic metals such as pure Ni, Fe, and Co, but also hybrid phases NPs, such as magnetic/no-magnetic HNPs to form multicomponent nanowires and nanotrusses.

On the other hand, the authors showed the crucial role of the magnetic field application to obtain these anisotropic nanostructures, as shown in figure 5.11. The assembly of Ni/NiO nanowires and nanostrusses occurs only when a magnet is attached to the substrate, where the NPs are deposited (see also figure 2.9) [37].

References

[1] Benelmekki M, Bohra M, Kim J-H, Diaz R E, Vernieres J, Grammatikopoulos P and Sowwan M 2014 Facile single-step synthesis of ternary multicore magneto-plasmonic nanoparticles *Nanoscale* **6** 3532–5

[2] David S, Polonschii C, Luculescu C, Gheorghiu M, Gáspár S and Gheorghiu E 2015 Magneto-plasmonic biosensor with enhanced analytical response and stability *Biosens. Bioelectron.* **63** 525–32

[3] Zou F, Zhou H, Tan T V, Kim J, Koh K and Lee J 2015 Dual-mode SERS-fluorescence immunoassay using graphene quantum dot labeling on one-dimensional aligned magneto-plasmonic nanoparticles *ACS Appl. Mater. Interfaces* **7** 12168–75

[4] Kim J-H and Benelmekki M 2016 Interfacial transformation of an amorphous carbon nanofilm upon Fe@Ag@Si nanoparticle landing and its colloidal nanoscrolls: enhanced nanocompositing-based performance for bioapplications *ACS Appl. Mater. Interfaces* **8** 33121–30

[5] Benelmekki M, Gasso S and Martinez Ll M 2020 Simultaneous optical and magnetophoretic monitoring of DNA hybridization using superparamagnetic and plasmonic colloids *J. Coll. Surf. Biointerfaces* **193** 111126

[6] Xia Y, Yang P, Sun Y, Wu Y, Mayers B, Gates B, Yin Y, Kim F and Yan H 2003 One-dimensional nanostructures: synthesis, characterization, and applications *Adv. Mater.* **15** 353–89

[7] Lisjak D and Mertelj A 2018 Anisotropic magnetic nanoparticles: a review of their properties, syntheses and potential applications *Prog. Mater. Sci.* **95** 286–328

[8] Benelmekki M 2015 *Designing Hybrid Nanoparticles IOP Concise Physics Series* (San Raphael, CA: Morgan & Claypool Publishers)

[9] Benelmekki M 2019 *Nanomaterials: The Original Product of Nanotechnology IOP Concise Physics Series* (San Raphael, CA: Morgan & Claypool Publishers)

[10] Yin Y and Alivisatos A 2005 Colloidal nanocrystal synthesis and the organic–inorganic interface *Nature* **437** 664–70

[11] Chen W *et al* 2019 Strong electronic interaction of amorphous Fe_2O_3 nanosheets with single-atom Pt toward enhanced carbon monoxide oxidation *Adv. Funct. Mater.* **29** 1904278

[12] Bao L, Low W-L, Jiang J and Ying J Y 2012 Colloidal synthesis of magnetic nanorods with tunable aspect ratios *J. Mater. Chem.* **22** 7117

[13] Meldrum F C and Cölfen H 2008 Controlling mineral morphologies and structures in biological and synthetic systems *Chem. Rev.* **108** 4332–432

[14] Moscoso-Londoño O, Muraca D, Tancredi P, Cosio-Castañeda C, Pirota K R and Socolovsky L M 2014 Physicochemical studies of complex silver-magnetite nanoheterodimers with controlled morphology *J. Phys. Chem.* C **118** 13168–76

[15] Yang H *et al* 2010 Porous α-Fe_2O_3 nanostructures with branched topology: growth, formation mechanism, and properties *CrystEngComm* **12** 1842–9

[16] Chen Y, Milenkovic S and Hassel A W 2008 Arrays of iso-oriented gold nanobelts *Nano Lett.* **8** 737–42

[17] Tran V T, Zhou H, Lee S, Hong S C, Kim J, Jeong S-Y and Lee J 2015 Magnetic-assembly mechanism of superparamagneto-plasmonic nanoparticles on a charged surface *ACS Appl. Mater. Interfaces* **7** 8650–8

[18] Palmer R E, Cai R and Vernieres J 2018 Synthesis without solvents: the cluster (nano-particle) beam route to catalysts and sensors *Acc. Chem. Res.* **51** 2296–304

[19] Benelmekki M, Singh V, Baughman K W, Bohra M and Kim J-H 2021 On the potentiality of a cluster-beam-source to produce free-standing one dimensional and two dimensional FeAg nanostructures: mechanisms underlying their shape genesis *Mater. Today Chem.* **19** 100405

[20] Wang Z L 1998 Structural analysis of self-assembling nanocrystal superlattices *Adv. Mater.* **10** 13–30

[21] Wu Z, Yang X and Wang Z 2019 Size effect on the spontaneous coalescence of nanowires *Nanotechnology* **30** 245601

[22] Cazaux J 2008 On some contrast reversals in SEM: application to metal/insulator systems *Ultramicroscopy* **108** 1645–52

[23] Benelmekki M, Vernieres J, Kim J-H, Diaz R E, Grammatikopoulos P and Sowwan M 2015 On the formation of ternary metallic-dielectric multicore-shell nanoparticles by inert-gas condensation method *Mater. Chem. Phys.* **151** 275–81

[24] Miranti R, Qayyuma M-S, Sharma A, Einarsrud M-A, Mestres N and Benelmekki M 2020 Spectroscopic study of partially oxidized BN nanoscrolls induced by low-frequency ultra-sonic irradiation *Appl. Surf. Sci.* **515** 146055

[25] Cao G and Wang Y 2011 *Nanostructures and Nanomaterials* 2nd edn (Singapore: World Scientific)

[26] Rapp B E 2017 The circular flow tube ed B E Rapp *Micro and Nano Technologies, Microfluidics: Modelling, Mechanics and Mathematics* (Amsterdam: Elsevier) ch 14 pp 309–13

[27] Beveridge J S, Stephens J R and Williams M E 2011 Differential magnetic catch and release: experimental parameters for controlled separation of magnetic nanoparticles *Analyst* **136** 2564–71

[28] Zborowski M, Sun L, Moore L R, Williams P S and Chalmers J J 1999 Continuous cell separation using novel magnetic quadrupole flow sorter *J. Magn. Magn. Mater.* **1999** 224–30

[29] Hummel R E 2001 Magnetic phenomena and their interpretation: classical approach *Electronic Properties of Materials* (Berlin: Springer))

[30] Andreu J S, Faraudo J, Camacho J, Benelmekki M, Rebollo C and Martinez Ll M 2011 Simple analytical model for the magnetophoretic separation of superparamagnetic dispersions in a uniform magnetic gradient *Phys. Rev.* E **84** 021402

[31] Faraudo J, Andreu J S, Calero C and Camacho J 2016 Predicting the self-assembly of superparamagnetic colloids under magnetic fields *Adv. Funct. Mater.* **26** 3837–58

[32] Bishop K J, Wilmer C E, Soh S and Grzybowski B A 2009 Nanoscale forces and their uses in self-assembly *Small* **5** 1600–30

[33] Wagle D V and Baker G A 2015 Cold welding: a phenomenon for spontaneous self-healing and shape genesis at the nanoscale *Mater. Horiz.* **2** 157–67

[34] Jose-Yacaman M, Gutierrez-Wing C, Miki M, Yang D-Q, Piyakis K N and Sacher E 2005 Surface diffusion and coalescence of mobile metal nanoparticles *J. Phys. Chem.* B **109** 9703–11

[35] Grammatikopoulos P, Cassidy C, Singh V, Benelmekki M and Sowwan M 2014 Coalescence behaviour of amorphous and crystalline tantalum nanoparticles: a molecular dynamics study *J. Mater. Sci.* **49** 3890–97

[36] Cassidy C, Singh V, Grammatikopoulos P, Djurabekova F and Nordlund K 2013 Inoculation of silicon nanoparticles with silver atoms *Sci. Rep.* **3** 3083

[37] Ekeroth S, Ikeda S, Boyd R, Münger P, Shimizu T and Helmersson U 2019 Impact of nanoparticle magnetization on the 3D formation of dual-phase Ni/NiO nanoparticle-based nanotrusses *J. Nanopart. Res.* **21** 228

[38] Ekeroth S, Münger E P, Boyd R, Ekspong J, Wågberg T, Edman L, Brenning N and Helmersson U 2018 Catalytic nanotruss structures realized by magnetic self-assembly in pulsed plasma *Nano Lett.* **18** 3132–7

IOP Publishing

Designing Hybrid Nanoparticles (Second Edition)

Maria Benelmekki

Chapter 6

Summary and future outlook

Substantial progress has been made in recent years designing and synthesizing multicomponent nanoparticles (NPs) that combine two or more component materials into one nanostructure. Such complex structures have the potential to combine magnetic, plasmonic, semiconducting and other physical or chemical properties into a single object, allowing an enhanced functionality. First efforts began with core–shell hybrid NPs, and then culminated in morphologies that deviate significantly from core–shell structures, such as heterodimers, or NPs decorated with several domains. Other emphases were put on the production of anisotropic nanostructures through the assembling of single-component or multicomponent NPs to obtain 1D, 2D and even 3D materials. In all cases, the enhanced and often new functionality resulting from the synergetic combination of parent properties, on one hand, and the anisotropic shapes on the other hand, allow expansion of the already wide applicability of hybrid NPs (HNPs).

HNPs in which spherical, rod or branched domains are connected in a controlled morphology are promising for a number of potential uses ranging from nanoelectronics to biomedical, photovoltaic and catalytic applications. For example: (1) magneto-semiconductor NPs are very interesting as bimodal tags offering the ability to visualize labelled cells using both magnetic resonance and fluorescence imaging techniques, while an external magnetic field may be employed for the directed assembly of such materials; (2) magneto-plasmonic NPs have been proposed for a large number for dual magneto-optical applications, theranostics, multimodal imaging, cell-sorting and biomagnetic separation; (3) metal–semiconductor heterodimers are ideal candidates for solar energy harvesting applications, and in particular photocatalysis where the semiconductor part of the NPs can be tailored for optimal light absorption, while the metal part facilitates charge separation.

Despite the wide variety of synthetic strategies that have been reported for the production of such advanced complex nanostructures, the development of new synthetic strategies that lead to a finer morphological control of NPs is still a challenging task. Beyond the control of NPs' morphology, one of the most important open questions is to understand how synthetic conditions can be controlled to combine several materials with different structural and physical properties.

Within this context, the magnetron-sputtering-based inert-gas condensation (MS-IGC) method is becoming one of the most widespread methods for the production of supported HNPs. This method allows accurate control on the morphology, chemical composition, and size distribution of the NPs. In addition, it offers the advantages to produce phase-separated NPs with free-residual contamination both at the interface between the different components and on the surface. As illustrated in figure 6.1, the controlled production of HNPs requires the integration of three fundamental parts: (1) the planning of the experimental conditions and procedures; (2) the choice of the characterization techniques at the nano scale; (3) a system-engineering capacity to optimize the instrument performances, towards large-scale applications. Chapters 3, 4 and 5 present examples of the design and planning of experimental conditions and the choice of characterization methods.

The first example addresses the synthesis and characterization of intermetallic FeAl NPs whose components are partially bulk-miscible. The advantage of working under vacuum, allows a better control of the NP oxidation process. The resulting NPs consisted of a nanoalloy of an iron aluminide crystalline core (DO_3 phase) encapsulated in an alumina shell that reduces inter-particle interactions and prevents further oxidation and segregation of the FeAl core. The second example addresses the synthesis of metallic-dielectric Ag–Si NPs, whose materials are bulk-immiscible. In this case, an appropriate tuning of the sputtering power ratio of the Si and Ag

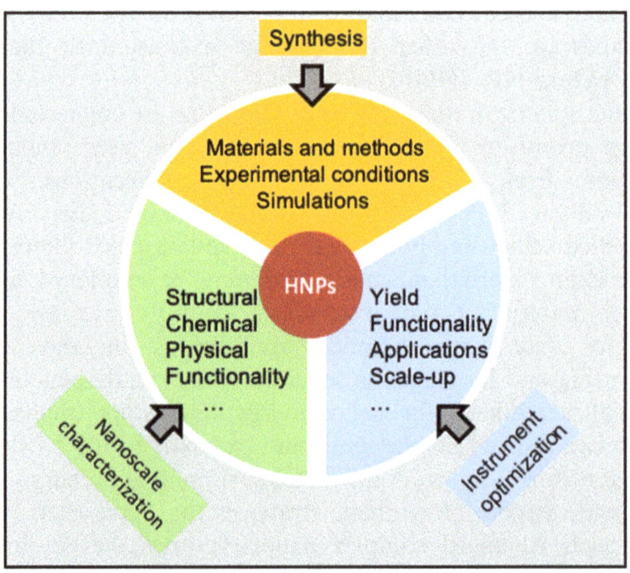

Figure 6.1. A diagram summarizing the key inputs for a controlled production of HNPs.

independent targets allowed the production of NPs with distinctive multicore–shell and core–satellite morphologies.

The third example consists of ternary NPs composed of magneto-plasmonic FeAg cores encapsulated with a Si shell. The size of the NPs and the shape of the cores in these NPs are modulated by controlling the deposition parameters as follows:

- Adjusting the discharge current on each target, allows the control of the plasma density on the surface of each target, and thus the size and crystallinity of the formed NPs.
- Reducing the differential pressure between the aggregation zone and the deposition chamber, increases the residence time of the NPs inside the aggregation zone enhancing the coalescence process, and therefore, larger NPs with a higher number of cores are formed.
- The immiscibility of Fe and Ag and the low surface energy of the amorphous Si play important roles in the position of each material within the hybrid NPs.

In addition, when these NPs are exposed to the atmosphere, the Si surface is oxidised to SiO_x, providing an easy path for their surface modification for a wide range of bioapplications.

The new chapter 5 presents examples of how a further modification of the conventional MS-IGC system leads to the production of one-dimensional and branched Fe, FeAg, and Ni/NiO nanostructures. The first example demonstrates that a balance between the magnetic force applied to the inflight Fe NPs on one hand, and the dragging force driven by the pressure differential between the different chambers of the MS-IGC, on the other hand, result in the formation of Fe and FeAg NRs. The example elucidates the mechanisms involved in the formation of the NRs and illustrates the importance of nanosize-effects when NPs are brought to intimate contact, through the interfacial transport of matter at the contact interfaces. It also shows the importance of the accurate choice and the correct use of the characterization techniques. Branched nanostructure formation is illustrated through the second example of this chapter. Fe and Ni/NiO nanowires and branched nanostructures are obtained by applying an external magnetic field to the deposition substrate. This example brings to mind the possibility of a controlled production of multicomponent magnetic/non-magnetic nanowires and branched nanostructures (nanotrusses).

However, for a better understanding of NPs' formation, further improvements in characterization methods of aggregation zone conditions are needed. So far, progress in the understanding of gas-phase HNPs has been achieved based on computational techniques. On the other hand, to make profitable the scale-up of this method, there is an urgent need to improve the yield and harvesting processes of the resulting HNPs. Chapter 2 discussed some of the efforts dedicated to system-engineering to improve the performances of conventional MS-IGC system. It also discussed the development of instruments based on new fundamentals such as the case of rare gas condensed matrices. Within this context, an example of Ag–TiO_2 and Au–TiO_2 HNPs obtained using MACS is presented in chapter 3. These HNPs showed catalytic activities comparable to those of catalysts made by conventional chemical methods, indicating the near-future potential of gas-phase systems in the field of heterogeneous catalysis.

Similarly to any other nanotechnology product, HNPs are facing challenges related to their large-scale production on one hand, and to their regulatory frameworks, which include health and safety at work, together with environmental protection, and management of the end-of-life of products, on the other hand. Within this context, there is a growing interest in the use of gas-phase-based-systems as an environmentally-friendly method to produce advanced HNPs, however, there are no studies available to evaluate the NP emission from a nanocluster source and to assess the risk in the working environment when using these systems. Therefore, the future will still witness important efforts in this research field.

www.ingramcontent.com/pod-product-compliance
Ingram Content Group UK Ltd.
Pitfield, Milton Keynes, MK11 3LW, UK
UKHW051848100926
8556IPUK00005B/519

9 780750 335850